My Little War
by John Martin

Edited By Rebecca Rutledge and Charles Martin.
Cover Image By David Woods.
Photos by John Martin.
My Little War Copyright © 2016 by John Martin. All rights reserved. To be printed in the United States of America. No part of this book may be used or reproduced in any manner whatsoever without written permission from the author except in the case of brief quotations embodied in critical articles and reviews. For information address Literati Press, 3010 Paseo Oklahoma City, OK 73103.

"A nation that draws too broad a difference between its scholars and its warriors has its thinking done by cowards and its fighting done by fools." —Thucydides

Editor's Note: This project began in 2003 when I pulled four cardboard boxes out of my father's closet. I began a long process of excavating hundreds of unsorted, yellowing, and White Out-smeared pages. These fractured memories from his time before, during, and after Vietnam revealed markers of post-traumatic stress syndrome and survivor guilt. We began a thirteen year process of cobbling those memories into this book. We chose a non-linear storytelling style because my father's journey from haunted soldier to family man was also not a simple straight line.

Introduction

It's okay now to display things like photos and medals. But I still don't like touching my uniform.

I still feel guilty about Staff Sergeant Son, Captain Thinh, and the other countless South Vietnamese that I met during my time in country. They trusted me and I let them down.

The faces have faded and the names are pretty much gone now. That is probably good. Mostly, I feel I couldn't have been the one who did the things detailed in this book. It had to have been someone else.

And then something will happen to bring back a memory in clear, real-time mental sharpness as bright as the moment it happened. For example, it took twenty years for me to realize why getting a haircut was pure living terror for me. For several years, the only person I would allow close to my hair was my wife, Teresa.

Getting a haircut from a barber was a white-knuckle affair. It took a real occasion before I would agree to one. Say a kid getting married.

The day my number-two son, Jeremy, got married, I was sitting in the barber's chair, talking to the barber, a young guy. Somehow the conversation turned to a review of the movie *Apocalypse Now*. He knew I was a Vietnam vet, and he wanted my impression of the movie. During the conversation I flashed back to something I had not thought about since Vietnam. I was thinking about the battalion barber.

The battalion barber was Vietnamese. My intel reports alerted us to a cell of spies on several of our compounds. Barbers. My barber was a spy. I got a haircut from him once a week, every week while I was in Vietnam. I suppose there is nothing special about getting a haircut from a spy.

He was easy to talk to.

Of course he knew I was the Intel Officer for the battalion. He knew I knew he was a spy. We used each other to our own ends. At least on the surface, everything was okay—that is, until he took that razor out and started slapping it on the leather strap to sharpen it. A part of me was always convinced that his next act was going to be to cut my throat. Man, I hated that slap, slap, slap. And I really hated him using that razor on my neck.

For a short time, he had my life right where he wanted it, once a week, every week.

Now, twenty years later, there I was, white knuckles and all, sitting in a barber chair listening to that slap, slap, slap.

I realized that my Vietnamese barber must have felt the same way I did every time he came through the front gate. He knew that I had a razor too. Slap, slap, slap.

And suddenly, getting a haircut wasn't quite so bad.

Last night, I had a dream that I had been recalled to active duty and ordered to take a medical team to Vietnam. I had a lot of problems to solve. Most important, I didn't have enough uniforms, and the uniforms I did have were not the right ones. I had to get transportation for the medical team. I had to get supplies. I was unable to get anything done, no matter how hard I tried.

Over the years, I have had several dreams about being sent back to Vietnam. Since 1985, I can remember three or four general themes. The first theme involved direct combat in an imaginary fire support base; this theme usually involved some sort of close-range fighting. Another theme involved flying in a Bird Dog aircraft and getting into some sort of scary situation with people on the ground trying to force us down.

The next theme involved me being sent back to Vietnam to find someone. It's not clear who, but it was someone like SSGT Son. In the dream, I find him living in a city kind of like what I remembered Long Binh looked like in 1970, but nicer.

He had a home and family, but we had nothing in common anymore.

The latest theme is like the dream I had last night. And I think I understand it. It is a combination of significant events in my life after 1970. In this dream, I stayed in the reserves for a long time. I never really had enough sets of uniforms to serve me right should I be recalled to active duty.

Not having enough of the right kind of uniforms always concerned me.

Happily, these dreams and memories are rare. I don't need them. I need memories of my grandkids.

SSG Son and CPT Thinh don't have memories of their grandkids. That could have been such a great gift.

In the early 1990s, I led my first church mission team to Slovakia. More recently, I led a medical mission team to Guatemala. In 2009, I participated in a construction mission to Vietnam.

Vietnam was a beautiful place, the Vietnamese were as special as I remembered them, and it was a great pleasure for me to associate with them again.

I broke down once while I was there. I cried uncontrollably, I think it was my way of saying goodbye to CPT Thinh and SSGT Son.

Or maybe I was trying to tell them I was sorry for not being the friend I should have been.

It all comes down to this: Why me? Why did I live through my eight months in Vietnam when so many other more deserving soldiers died? Why was I then blessed with a support system strong enough to hold me together during my own little war, a war that has never really ended and probably never will?

I know that I am lucky, and I pray that I will one day understand why I was saved.

Finally, it has always confused me that for my Army's entire Vietnam experience, we won every battle we fought. But we lost the war. A country and friends were destroyed. I wondered how that could be.

It took me a long time to realize that wars are not won with bullets and bombs. Wars are won by dedicated people with ideas, bold ideas that have the power to change minds. Bullets and bombs might get someone's attention, but that is not enough to change their mind. A visionary is needed to provide a reason for change and an opportunity to change and the road to win the war. It is my opinion that if we had more visionaries than simpletons in Vietnam in 1970, we could have won that war by 1972. You will meet some of each in this story. For obvious reasons, to tell this story I have changed some names; but, you will meet visionaries like Doc Carr and SSG Son, CPT Thinh and Lieutenant Colonel

Mullens. You will also meet some simpletons, like Lieutenant Colonel Gray and Major Douche. There are more visionaries in my story than there are simpletons; because, I want it to be a good story.

My story is not typical of the stories of the U.S. Army in South Vietnam in 1970. Except for that brief period when the Army was on the attack in Cambodia, the mission was to defend and survive until the freedom birds took the Army home. My Army was lousy at defense. A good, passive defense takes waiting, but waiting often leads to malaise which leads to a loss of focus on the mission. Absent an inspired leadership in My Army in South Vietnam, there was a breakdown of esprit de corps and moral structure. This gave way to black markets, drug issues, and inflamed cultural and racial divides. Sometimes outright rebellion.

But, then, what do I know?

I think I have been a fool.

Saigon
September 1970

My tour of duty in Vietnam was almost over. I would reach my separation date in mid-September, and considering my experiences thus far, I was not interested in staying in the Army. Since starting my tour in Vietnam, I had not taken any R and R. I hadn't even taken a day off.

As Staff Sergeant Son and I were returning from a trip in my Jeep to somewhere in the Long Binh/Bien Hoa area of South Vietnam, I mentioned to him that I wished I could take some time just to see the other side of Vietnam. He thought it over and then asked me if I would like to meet his family.

Son's parents lived in Saigon. His father was a civil servant who worked for the city. Son told me his father had spent his adult life as a civil servant. His career had started in Hanoi, where he had ties to the French back in the early 1950's. When the French lost control of French Indochina, Mr. Son took his family south to Saigon and became a civil servant in the new country of South Vietnam.

Son had been a little boy when his family moved to Saigon. He told me about his memories of that trip. Had I had to make a trip like that as a little boy, I'm sure it would have scared the crap out of me.

I told him I'd be honored to meet his parents. He said I'd be meeting his grandparents, too, since they all lived together. .

"I think that would be even better, Sergeant," I said. "I would like to see for myself how many generations it took to put a cowboy like you together."

"Fuck you, *Diwi*."

I had used the term cowboy to describe Son on other occasions. He understood that when I used it I was giving him a backhanded compliment. South Vietnam had cowboys in 1970. They were not cowboys as we understood a cowboy to be. They were a gang of criminals who were dangerous characters. Son understood that when I called him a cowboy, I was using an American cowboy as my reference. He knew about John Wayne. When I told him I thought of him as short John Wayne riding his horse into a bunch of wild Indians to save a bunch of farmers holed up in a circle of wagons, he liked that comparison. But then he'd say, "Fuck you, *Diwi*" because he liked to say it.

I asked our battalion executive officer, Major Major, for a 24-hour pass, and he granted it. Sergeant Son drove the Jeep to Saigon to meet his family. We were to be at his home for supper. Son suggested that after supper he would take me downtown and show me the nightlife. I told him that I'd prefer to stay with his family and avoid Americans as much as possible.

I was very nervous about meeting Son's family. So I pumped him on the proper manners I should use while there. I did not want to insult anyone just because I did not know something. He explained that his father spoke English, but that neither his mother nor his grandparents did. He said that he and his father would translate for me.

I asked Son if it would be appropriate for me to give gifts to his parents. He said his father liked Johnny Walker Red and his mother liked candy.

"But it really isn't necessary," he said, "since you helped get the refrigerator." The refrigerator had allowed his mother to stock up on food. The magic of ice was a great gift, a significant saver of time for her. She no longer had to go shopping for perishable food for every meal. She could use that time for other things now.

I couldn't find any good chocolate in the PX (Post Exchange) at Long Binh, so I decided to get several different kinds of candy bars and an assortment of gum and other junk food. The quart of Johnny Walker Red was easy, of course. Getting the presents wrapped was another matter until Son suggested that we get it done in the village. Such a simple solution. After all this time in Vietnam, after all the time spent with Son, and I still did not think like I was in Vietnam. I was still thinking like an idiot American.

When we arrived at the Son residence, my first impression was how clean and neat the small yard was. I was surprised to find the home was about the same size as the maternity ward we had just finished rebuilding as a Civic Action Project. His home was small for American standards. It was very clean. The floor was a beautiful ceramic tile. The walls and ceiling were white. There were pictures and personal family things. The furniture was tiny, made out of highly polished teak wood.

Son introduced me to his grandparents and his parents. I gave Mr. Son his present and turned to Mrs. Son and Grandmother Son. I told Staff Sergeant Son to tell his mother and grandmother there was a story that went with their present, and that I would like to tell it to them before they opened it.

First I had Son ask his mother and grandmother if they'd ever heard of Boggy Bait. They hadn't.

"Ask them if they can say 'Boggy Bait,'" I said to Son.

They tried, and we laughed.

I went on. "In America, when you join the Army, the first thing you lose is your hair. The second thing you lose is the privilege of eating Boggy Bait. It is not so bad losing your hair, since everyone around you has also lost their hair. After a while, it grows back, no sweat. But not being able to eat Boggy Bait will drive you crazy. Numma Ten!"

I looked at Son. "Let's try to say 'Boggy Bait' again."

Again, we laughed.

"After about the second week of being an American soldier without any Boggy Bait, you are ready to fight a tiger, barehanded, with one arm tied behind your back and a blindfold over your eyes."

As Son translated, he picked up that I was pulling his mom's and grandmother's legs, and he warmed to the task. I got up and put one arm behind my back, closed my eyes, and growled. We all laughed again.

"These American soldiers are so bad that there is a law in the United States that says that an American soldier without Boggy Bait must be kept away from all American women for a period of 48 hours. This law was pushed through the American Congress by the fathers of American girls. The lack of Boggy Bait makes young American soldiers act and smell funny."

To that last statement they all agreed. Americans really do smell funny.

"For an American soldier to be without Boggy Bait is so serious that our Congress passed a law that says no women can be in the Army. We are afraid that if American women went without Boggy Bait very long, they would beat up American men."

That seemed to make sense to them. Why else were there no American women in the Army?

Son winked at me and said, "Mother and Grandmother know women were better than men anyway."

"So that you can understand what makes us Americans so strange, I have brought you a present of Boggy Bait. Please open it now, and I will be happy to explain how it works."

Grandmother, Grandfather, and Mr. Son watched closely as Mrs. Son unwrapped the box with tentative fingers. Since Son knew what was in the box, he would respond in Vietnamese to their whispered questions. He set them up perfectly.

When she saw all the candy bars, she burst out laughing. Son relaxed considerably. He knew his family was going to accept this crazy American officer.

Considering the fact that I am not a fan of Johnny Walker Red or bubble gum, the combination of both in my mouth at the same time was not the tastiest experience for me. But you gotta do what you gotta do. And what the hell, Son's family really got a kick out of it the first time I blew a bubble and it popped and left gum in my mustache. I got a kick out of watching them try to blow bubbles. The betel nut tooth-stained smiles punctuated with bubble gum were a picture to behold.

They didn't have any problem with Johnny Walker and bubble gum.

Maybe it was my mood, maybe it was the atmosphere, but surely it was the most wonderful food I'd ever eaten. Son said I was not to worry about eating too much. They would want me to eat the very last bit of food if I could. So I did. There was fish, shrimp, and indecent, naked chicken. I never realized that those featherless little pieces of meat running around in front of God and everybody could taste so good.

After the meal, they gave me a drink of a sweet, fruity liqueur with a powerful alcohol kick to it. I'd never tried the drink before. When I asked Son what it was, he said I didn't want to know, and that he would tell me some other time. So I took him at his word and enjoyed myself. He never told me what it was.

As I relaxed, I began to take inventory of my surroundings. I noticed a shrine on the other side of the room and asked Son about it. He explained that the shrine was for his brother who had been killed in 1969. His brother was a ranger and had died while on a special mission in Cambodia.

Slowly the mood began to change. We talked about the war and Vietnamization. Then Grandfather Son asked me if I really believed that an American had walked on the moon.

"Yes, sir, I do," I said. "While I was in Germany, I listened to the communications with the astronauts with special radios that we had."

Grandfather Son said that it was impossible, it could not be done and that the whole thing was just American propaganda.

I had never thought of it as propaganda before. I knew that it was real and until that moment I had equated propaganda with lies. Suddenly, I realized that what was real to me may not be real to them. I did not want to offend him, so I chose my answer carefully.

"Yes, you are right, it is propaganda. But I believe that this propaganda is true. I have no doubt that Americans have walked on the moon."

"It is impossible, it cannot be done. It is just propaganda."

"Sir, can I ask you to ask your grandson what he thinks?"

After Son spoke to his grandfather, I asked him what they had talked about and what he thought. His answer was, "I don't know."

This young man was my friend. He would and had risked his life for and with me. But he thought I might be lying to him, and there was nothing I could do or say to make him believe in me. At that moment, I desperately wished I believed in God because I could have really used His help.

I looked at Grandfather and said, through Son, "Sir, if you are right, then it means that my country has lied to both of us. I do not believe that they lied to me, but there is no way I can prove to you that I am right."

There was no response.

Mr. Son asked me to explain what happened at Kent State University.

"As far as I know, a group of students were demonstrating against our invasion of Cambodia. Something happened that caused American soldiers to open fire on the demonstrators and a few were killed."

"Why were they demonstrating against us? We are your friends."

"There are those in America who do not believe in what we are doing in South Vietnam. Either they are afraid of the risk to their security if they were forced to fight for your freedom, or they believe what Hanoi is telling them."

"What do you believe?"

"Do you want me to tell you the official line, or do you want me to tell you what is in my heart?"

"I want to know what you believe."

"I have never lived where a foreign government dominated all. I do not know what it would be like not to have the freedom to think and speak what I feel. My ancestors lived under the domination of the British, the French and the Spanish. Almost two hundred years ago my ancestors were able to throw off foreign rule and domination. Since then we have been our own masters. We have made some terrible mistakes along the way; but, the mistakes were ours. All in all, we have been successful as a nation."

"For me, as an American, it means that I have rights to personal freedoms that do not exist other places. For me, as an individual, I want to give you those same freedoms."

"What do you mean by "personal freedoms'?"

"Personal freedoms are the rights to experience the world in your own way, to speak out without fear or repression, to choose to be a leader or a follower without the fear of abuse of power."

"That is impossible."

"Where I grew up it is not impossible. In my heart, I believe that when this war is over, there will be a time when you or your children will be able to experience life without American soldiers, without military vehicles shoving you off the road, without military jet aircraft overhead. Someday, if you see a soldier, he will have on an ARVN uniform like the one your son wears. But, I hope that someday, you would not see a soldier at all. Most of all, I believe that someday you will be able to control your own destiny without interference from anyone outside your country, including people like me. I believe this so much that I am willing to die for it."

He looked at me for a moment, shook his head, and said, "Captain, I hope that you are not offended if I say that I do not feel as you do. There will never be a time when I or my children will experience what you call 'personal freedom.' I do not understand what you are talking about, and I do not believe I want what you are talking about. It would be enough for me that you not die for what you believe, but that you just be there when I need you."

"Fair enough."

"Then let us drink to our friendship. But before we do, please explain to me how something like what happened at Kent State could happen. I do not understand. Please explain how Americans can openly support Hanoi, your enemy, and speak out against what you are doing here and be treated as heroes? How can you say that you will be here when we need you when all of these others are saying the opposite?"

"The people who are doing what you say are a minority in my country and they do not have a lot of power."

"If that is so, then why do I know about them?"

"I don't know, sir. I can only say that I believe that America would not let such an important ally down in a time of need. We never have and I believe we never will."

"I hope you are right, Captain."

A cold shiver ran down my spine as I thought about what a terrible cost he would pay if I was wrong.

Later, Son and I excused ourselves and went for a walk through the village. We ended up sitting by the river, listening to the sounds of the night. I could hear the distant sound of the motor of a sampan moving down the river, the sound of waves slapping on the riverbank. I could hear varmints moving through the dead leaves on the ground. It was a peaceful evening.

We sat there for several minutes not speaking. I absentmindedly picked up pieces of dirt and threw them into the water.

"What are you thinking about, *Diwi*?"

"I don't know—your family, I guess."

"Do you like my family?"

"Yes, Tran. I do. Thanks for bringing me here."

"No sweat, *Diwi*."

Off to our left, I heard the sounds of a family preparing for the night, furniture moving on the floor, doors and shutters closing, unconcerned conversations and kids being scolded. They were not preparing for the war. They were preparing for the peace. Just like home. There were lights in houses across the river. Just like home.

The stars were out and they were crystal clear. Just like home. It was the first moment of real peace I had known in months.

Beauty still existed in the world.

"Tran?"

"Yes, *Diwi*?"

"This is what I want for your grandfather."

I laid back and gazed up into the heavens. The stars were so clear. After a few moments of listening and dreaming, I began to feel a presence. It was as if I were being gathered up, not by physical hands, but by an idea, a revelation.

Son asked me, "*Diwi*, why don't you ever talk about your wife?"

"What?"

"Why don't you ever talk about your wife?"

"Hell, I don't know. I guess because I don't ever think about her."

"My grandfather has a story about women. You want to hear it?"

"Yeah."

"He says that there are three perfect women for every man. The first woman the man loses because he does not know what is in his heart. The second woman he loses because he does not know what is in her heart. He loses the third because he believes there is nothing left in his heart."

"What are you saying Tran?"

"I am concerned for you, *Diwi*."

"I guess I don't talk about her because I don't think about her. I don't think about her because I don't feel she is a part of all this. I guess, maybe, I resent that she is not a part of this. She should be. Since she is not a part of this, she will never understand. It's as if we never had anything in common."

After a moment, I added: "Isn't it strange, I can't even remember what she looks like?"

"*Diwi*, for you it will be over soon."

"I hope so," I said. "Let's go in."

That night, I woke up screaming from a recurring dream I'd been having about a rat. It had got me bad this time. I dreamed the rat was eating my face and I couldn't feel anything, nor could I stop it.

"It's okay," Son said to me. "Everyone has bad dreams."

The next day as we drove back to the battalion, I asked myself over and over again:

"*Am I right? Am I right?*"

Ft. Sill, Oklahoma
1972

Captain Thinh thought getting money from a bank ATM was a hoot.

He and a handful of South Vietnamese officers were at Fort Sill to learn advanced military leadership and tactical skills. The training was to prepare them for promotion to field grade rank. These officers were identified for high leadership roles in South Vietnam for the defense of their country.

Thinh and I shared officer quarters while both of us attended the Field Artillery Officers' Advanced Training Course. We came to share a friendship. He learned about my family and I learned about his. As time passed, he became more and more enamored with the American way of life. He told me once, "You Americans have machines for everything," including machines that spit out cash on demand.

Here's a funny story about Captain Thinh: It was February, 1973. We were outside on one of those bone-chilling Oklahoma days, black ice everywhere, getting trained on adjusting artillery fire. For several hours at a time, we stood on top of a hill with field glasses and watched artillery rounds explode. The cold slowly crept into our clothes and brought a real chill into our bodies. Thinh had never been truly cold. As his body temperature dropped, he started shaking. As he got colder, he lost feeling in his fingers and toes. Before long he started having trouble walking, talking, and even using things like his binoculars. We were watching artillery rounds hit the ground when he told me he thought he was dying. He wasn't joking. I turned away so he wouldn't see me smile. I knew he was okay. I was cold, too.

But not that cold.

Back in my car at the end of the day, he begged me to take him to the hospital. He was sure he was dying. I said I'd take him, but only after he took a shower. I told him to make the water as hot as he could stand it, and to stay in the hot shower for twenty or thirty minutes. Then, if he didn't feel better, I'd take him to the hospital.

I also told him that after the hot shower, he'd be able to go outside with no clothes on and not feel the cold. He thought I was crazy, so I made a bet with him. Since I drove him everywhere he wanted to go, he always insisted on being the one to clean the snow and ice off the windshield of the car. I bet him a steak dinner that if he took a hot shower the next time he needed to scrape off the windshield, he'd be able to do it with nothing on but shorts.

He took me up on it.

I won the bet. We drove into Lawton and consumed a steak at his expense.

From that point on, neither he nor the other Vietnamese officers in the class had a problem with the cold. They enjoyed the snow just like I did when I was young.

Thinh was a typical Vietnamese family man. He loved his wife and kids and he missed them. Every week or so I got the privilege of viewing new family photographs and listening to some new story about the trials of being a dad. He was short, not much over five feet tall, strong and in reasonably good shape. I had no illusions that he could, if he wanted to, kick my butt. I did not intend to find out.

We had a lot in common. We had our differences.

The biggest difference between Thinh and me was his determination to serve his country, no matter what it cost him. That determination was something I'd once felt. I envied him for it.

One Saturday afternoon, we were sitting around enjoying a beer when he told me he'd decided to mortgage his home in Saigon so he could run for political office. He felt a calling, he said, to be more involved in the future of his country. My first reaction was to wonder how he could be more involved than he was right now. I was flabbergasted, and I told him so. I reminded him there was a real chance that he'd fail to get elected. I warned him not to put his family at such risk. He said it didn't make any difference; he had to do it.

Now, *that* was patriotism I thought. What a guy. I wished I could feel that way about something, anything.

In 1972 and 1973, Nixon had been caught lying about Watergate and his minions were in Paris trying to negotiate a peace settlement with North Vietnam. Thinh and a few of the other Vietnamese officers would sit down with me often to discuss these things. They were troubled about the peace negotiations.

Over and over, they told me they suspected the U.S. would abandon them and their cause. I always responded that America had never abandoned an ally, and that we wouldn't abandon them. I believed what I told them.

Thinh said he feared America was uncivilized. That was the word he kept using. He said Americans were more concerned about machines than life. He was afraid we couldn't be trusted. That my friend would say that to me kind of hurt.

My heritage wouldn't let me believe him. America would never break a promise to an ally. Running away was unthinkable. The president, my commander in chief, had committed me and the rest of us to them. That was enough for me.

But Thinh persisted. He feared that we Americans didn't understand Vietnam or the concepts of face and friendship. We spent long hours discussing America, South Vietnam and America's claim of the "friendship" between us, but Thinh remained dubious.

"Your word is your bond," he tried to explain to me. "Your word and your bond are your honor."

For him, honor and civilization were one and the same. Over and over, he voiced his fear that America was uncivilized. Over and over, I tried to assure him we'd stick by our commitment to what he and his people were trying to do.

I knew that nothing I said would convince him. But I also knew that, when and if the time came, America would support South Vietnam, whatever the cost. We'd just have to show him by deed, not by rhetoric.

Home
New Year's Eve 1972

I didn't want to give my ex-wife a name. I can't explain why.

As we neared publication, I still didn't want to give her a name, but I did. Her name is Twinkle Toes. Twinky for short. I never actually called her that and, to my knowledge, no one has ever called her that.

But now she has a name because I suppose she deserves one.

—

I was in Wyoming on Christmas leave from Ft. Sill, where I'd left Thinh and the other Vietnamese officers. Twinky had dismissed the idea of following me to Ft. Sill. Her career was more important. Other than missing my daughter, Anna, I had no problem with the split household. It was nothing new. We'd lived apart since 1969.

It was late morning when I arrived at the house in Rawlins. Twinky was at work and Anna was at daycare. I planned to unload the car and rest a little, then go see my parents.

Hot, spoiled air blasted me as I opened the front door. It smelled like old food and garbage. The place was a wreck, dirty clothes and trash all over the place. The trash hadn't been taken out in weeks. A thick layer of dust covered all the furniture.

Pissed me off.

So I spent the rest of the day cleaning house. I got madder and madder as I worked.

I dumped Twinky's clothes in the washer and started washing them. While the washer was running, I separated Anna's clothes by color and took a little more care washing them. Then I turned my attention to the kitchen.

The dried food on the dishes was as hard as concrete. I worked up a sweat scraping and hand washing them. It took a long time.

The longer it took, the madder I got.

There was mail, newspapers, and National Organization of Women magazines everywhere. I gathered up the mail and put it in a pile on the kitchen table. I threw all the magazines in the trash.

Around six that evening, the front door opened and Anna came running in. She jumped into my arms yelling: "Daddy, Daddy, Daddy!"

I picked her up and gave her a big hug and whispered into her ear "Now this is what I need from my little lady!"

Twinky came in behind Anna. She looked around, closed the door, took off her coat, and walked into the bedroom.

In a couple of minutes she was back.

"What did you do with my magazines?" she said.

"I threw them away."

"You did what?"

"I threw them away."

"They're mine. Why did you do that?"

"Well, this is my home, too. I figured that if they were all that important to you, you'd put them up."

Twinky turned on her heel and left the room, leaving Anna and me alone to play.

A couple days after Christmas, Twinky's parents picked up Anna so she could stay with them while Twinky and I celebrated the New Year.

New Year's Eve turned out to be a blizzard kind of night. Not just a friendly blizzard, but a blizzard designed to get your attention. A stay inside kind of blizzard.

Just right for some serious bar hopping. Any excuse for us not to be alone with each other. Twinky still hadn't forgiven me for throwing away her modern-age literature, and I hadn't forgotten what the house had looked like when I got home, so neither of us were all that anxious to spend the evening with each other. The thought of spending New Year's Eve alone with Twinky did not sit well with me. Twinky was having the same concern about me. Spending some time in public made sense to both of us. So we crawled into the car and I drove through the blizzard to Front Street, where I parked in front of the Silver Spur.

We took our place at a table in the Silver Spur and ordered the first round of drinks. Neither of us had much to say, so we just sat there and ordered more Canadian Clubs and Black Russians. After an hour or so, we felt a little more relaxed. Slightly inebriated is a better description.

A young couple came into the bar and took the table next to us. The lady caught my attention. Her date helped her off with her coat and held her chair as she sat. She had short blond hair and wore a black dress that was filled out correctly. He took great care folding her coat over the back of another chair, then he sat down and took her hands in his.

In a very low tone just loud enough for me to hear, Twinky said, "That is disgusting!"

"What?"

"That guy is treating her like he owns her, and she's letting him do it."

"What would you have her do, burn her bra?"

I turned around in my chair and addressed the young man. "My wife thinks you are an asshole." I looked at the young lady and said, "She thinks you are an idiot."

I turned back to Twinky. "There," I said. "Feel better now?"

"You are a son-of-a-bitch."

"Yeah, I'm a son-of-a-bitch. But I'm your son-of-a-bitch, and you'd be wise not to forget that."

There was fire in her eyes. "Fuck you."

"Nice talk!" I said. I thought about adding "garbage mouth"; but, I thought better of it.

I looked at Twinky—no, I stared at her. *What the Hell am I doing here,* I wondered.

Twinky stared back at me. I saw rebellion in her eyes.

As I sat there and stared at her, I began to think about things long suppressed. Suddenly, all my frustration over this marriage burst forth. Things I had tucked away back in the back of my mind with the label "Shut Up Subjects."

The Shut Up Subjects were too raw and dangerous for me to deal with even in a marginal way. Twinky was responsible for several of these subjects. She had forced a lot of crap on me from the day I decided to let her use my name.

I remembered the letter she wrote to me while I was in Vietnam. That letter told me not to write to her anymore because my letters were too depressing for her. That letter hurt.

As I looked at her, at all that self-righteous indignation on her face, I made a barely audible comment, more to myself than to her. "You're right about one thing," I said. "This is depressing."

"What did you say?"

"Nothing important."

My mind moved from the letter to my memories of my first night back home from Vietnam. My memory about how it felt when

I finally arrived in Salt Lake aboard a United Airlines flight from San Francisco. I exited the plane expecting to see Twinky waiting for me. She was not there. I stood there at the gate, alone, not knowing what to do next. Maybe I was not supposed to come home. That feeling hurt.

The memories of our first night together when she made it clear she thought I was unclean. That night hurt, a lot.

I swirled my drink and mused, "I just realized I never did clean up good enough for you, did I?"

"What are you talking about?"

"Oh, just thinking back."

Then my mind moved to Roger. Sometime in the spring of 1968, Twinky confessed to me that Roger was Anna's natural father. Roger was the young stud who did not stand up and be responsible for his grazing in her grass. Roger was the asshole who was okay with me cleaning up his cum.

I looked at Twinky, finished off my Canadian Club, and said, "Roger."

She blinked.

My mind then shifted to the conversation we had had about Roger. Sometime in 1971, Twinky asked me if it was okay if Roger visited Anna. I couldn't believe she was dumb enough to think I would agree to such a stupid thing. I had just gone through hell in Vietnam, was still going through hell for her, and this is how she thought? I responded to her, "If Roger gets anywhere near Anna, I will kill him."

She knew I meant it.

Later, Twinky told me that Roger had something wrong inside his head and that it might be terminal. I couldn't believe it—Roger was not going to go away. Or maybe he was. Couldn't happen to a nicer guy. I responded with a laugh and said something to the effect that there really was a God.

Roger had come up again a couple of nights before, when Twinky notified me that she, Roger, and some other state employees were going to attend a two-week training seminar in January and they had decided to rent a house to live in.

I ordered another round of drinks.

"So Roger never did kick the bucket. Too bad."

"Roger is none of your business!"

My mind would not leave Roger. I thought back to the day in October 1970 when Twinky and I traveled from Rawlins to Denver so I could apply for a job with the Secret Service. To get to Denver, we had to drive through Laramie. As we approached Laramie, Twinky suggested we stop for a while. She wanted to call some friends at the University of Wyoming.

I agreed.

We stopped for lunch and she made her calls. When she returned to the table she suggested that we go to the university before we headed out for Denver. She would meet her friends at the Student Union.

I agreed.

I parked our car in front of the Union on campus. I told her that I wanted to go to the bookstore in the Union. She said she would wait for me outside. So I went in. As I walked through a door to the Union, my antennae went up. I smelled a rat.

Instead of going to the bookstore, I stopped just inside the Union door and watched her.

In a couple of minutes a guy walked up to her and they started talking. She reached out and took his hand. She moved in to him so that their bodies were touching. When they were finished talking she reached up and kissed him on the lips.

So this must be Anna's father, I thought. While I was in Vietnam and she was finishing up her bachelor's degree at Wyoming, she was spending my combat pay so this asshole could dip his wick between her big toes.

I was startled out of my thoughts by the cocktail waitress as she sat our round of drinks on the table. As I looked into my drink and swirled it I thought, *Roger.* I remembered being really pissed off back then. I was fucking furious now.

"So you think Roger is none of my business," I said. "I don't think so."

Sitting at that table in the Silver Spur, looking at this person who had just said "fuck you" to me and who had been saying "fuck you" to me one way or another for two years made me realize it was time to make a decision.

Up to this moment I'd been willing to look the other way because Twinky was the mother of my daughter, who I dearly loved. For two years, because of Anna, I looked the other way as Twinky insulted me and disrespected my name.

It was time to bring this brouhaha to a head. I hadn't been expecting it to explode in a bar on Front Street in Rawlins, Wyoming on New Year's Eve; but, here we were.

I looked at her and shook my head in disgust. "To think I once considered you a friend."

She yelled, "You don't know anything!"

"I know this much, the next time an artillery round hits next to me and I have to pick up a severed hand which is all that is left of what used to be an American soldier, if the fingernails on that hand have nail polish on, I'll be ready to think equality and respect."

Then I added, "Until that happens, I'll thank you to keep that hole in your face shut!"

She blinked again. Her mouth opened, but nothing came out. I had never said anything like that to her before. I couldn't tell if I had scared her or if she was shocked that I would be so crass.

I expected her to say something. I wanted her to say something, anything. Finally she blurted, "You think you are some kind of a man because you went to Vietnam?"

"What the hell do you know about a man? I know this much, while I was being shot at, you strapped a mattress to your back and let Roger shoot wet and sticky bullets between your legs." Then I added: "That don't get you my respect!"

She screamed, "Fuck you!"

"Thanks for the offer, but I have been there. It's not all that great. I'll pass."

She stood up and threw her drink in my face. I saw the drink coming and I moved my head to the side. Most of it landed on the floor. She slammed the drink glass down on the table, put on her coat, and stomped out of the bar.

I turned in my seat to look at the young couple. "I am real sorry you had to witness that," I said. Then I smiled. "I guess we're leaving now. Happy New Year."

I got up, put on my coat, and walked toward the door of the bar. Everyone was quiet, looking at me.

Probably made their evening, I thought. Most of them were friends of ours. Some of them were close friends. They probably knew more about what had been going on than I did.

Stepping out into the cold Wyoming night, I took a deep breath of the snow-driven air. It hadn't been so bad, I thought.

Actually, it felt pretty damned good.

When we got home, Twinky went to the bedroom and I went to the couch.

Today, I have no idea when the conflict that destroyed our marriage started. I don't know when she turned against me, nor do I know why. I suspect she was against me from the beginning. By giving her and her daughter my name, I may have created a debt that she was unable, in her liberated mind, to pay, and she resented me for it. Or maybe it was me. Maybe I never considered us to be really married in the first place.

Roger was not the cause of her resentment. He was the tool she used to damage any hope we might have had for happiness.

Now that I think about it, our argument might have started in June of 1968, when she asked me for advice and I responded by asking her to marry me for the sake of her daughter. But in 1972, the important point was the argument in the bar made it clear that our marriage was finished. She wanted to be free of the tyranny of marriage to me. I was finished being her husband.

The next day, the first of January, 1973, I called my parents to let them know I was heading back to Ft. Sill. Then I packed my bags, loaded them into my car, backed the car into the snow-drifted street, and started busting my way toward the highway. The thousand-mile trip to Oklahoma, much of it in bad winter weather, gave me time to sort things out. I remembered the disgust

I felt for Twinky during that argument in the bar. It was unsettling. Had I contemplated slapping her? Yes, maybe I had. Had I thought about doing something more severe?

It was a question to ponder.

It scared me that I might raise my hand against my wife. My father had taught me better than that.

The snowplows had busted the big snow drifts, but the wind was building new ones on the road surface that was already covered with packed snow and ice. One thought lead to another as I drove. I thought about what I had to do for Anna. She was the one good, pure relationship in my marriage. She gave me a reason to keep going. She lived with her mother, not me. For large periods of her life I had not been around. It was clear to me that I was not prepared to be her day-to-day anchor. It was also clear to me that I could no longer be around her and her mother at the same time.

After a few miles of driving, I began to realize that my love for my daughter and my disgust for my wife were diametrically opposed, irreconcilable feelings that would cause my daughter damage.

I really did not like my wife, Anna's mother, and that dislike had real risk. I had just insulted her in public. Until this moment, I had not thought about physical violence against her. But I was thinking about it now. Just how far would I take thoughts like this?

This line of thinking was not good, and it was not right. If I continued this charade with Twinky, my daughter could end up the real loser. Anna could end up with no mother and no father.

That was unacceptable.

As I drove down that long, lonely highway, I had to force my eyes to look through the wind-driven snow so I could try to keep the car traveling in the tire tracks of vehicles ahead of me. I had to keep the car traveling slow enough to be able to stop but fast enough to bust through the snow banks. At times I could see several hundred feet ahead of me. At times I could not see past the hood of the car. It was slow going.

Somewhere between Laramie, Wyoming and Ft. Collins, Colorado, I made the decision that it was time for me to walk away from my marriage and my family.

I had done what I had started out to do back in June of 1968. I gave that baby girl my name, and I had paid the price for that decision. Mission accomplished. Now, the only way I could protect Anna was to not be part of her life any longer.

I looked out the windshield at the white world in front of me and smiled. I knew this was right.

Within a year, I was still a daddy on paper, but I was no longer a husband. A few years later Anna asked me if it was okay if she stopped using my name. She didn't want me to be her daddy anymore. I said okay.

There was no need to get upset or to fight about it. There was a lot of her mother in her. That was something I could never change. For a very important part of her life, my name had made Anna respectable in the eyes of the world, and that was all that really mattered.

Lawton, Oklahoma
Easter 1975

In 1975, I was working on an MBA at the University of Wyoming. My girlfriend, Teresa, was living in Lawton with her two sons—five-year-old Matt and three-year-old Jeremy.

I had met Teresa while I was going to school in Ft. Sill in Lawton about a year earlier. She'd been mad as hell at my friend J.B., who was dating her best friend. He had been caught with a go-go dancer/call girl. Teresa's best friend was devastated that J.B. would do something like that. Teresa was pissed at J.B. for treating her best friend so shabbily. It became my job to protect J.B. from Teresa. J.B. was a good friend and he was scared of Teresa.

Teresa was drop-jaw beautiful. She had red hair that was cut short and a body that required my constant attention. I couldn't stop looking at her. Somewhere between defending J.B. and my MBA, I fell head over heels in love with her.

I appreciated and still appreciate what J.B. did for me. Bless his heart.

Teresa would write me two or three letters a week. I read each one over and over. Her letters made *Playboy* seem like a first-grade reader. Not that I read a lot of articles in *Playboy*, mind you. I was more of a pictures kind of guy.

Naturally, a long-distance phone call would follow a letter. And then it was phone sex for an hour or so. It got so bad that one of my weekly paychecks went to pay for my phone bill each month.

It must have been about ten thirty when I called her one night in February. I could tell she had something on her mind.

She cut to the chase. "Are you going to marry me or not? I have received two other proposals and I have to make a decision now."

I was shocked. My mind was having trouble understanding what she had said. Did I just hear her ask me to marry her? How in the hell could I do that? I was living hand to mouth, earning just enough money to pay for college. How could I afford a family now? And who in the hell were these other guys?

I said, "Marrying you is my dream. Marrying you is why I am going to college. But right now, I have nothing to give you. I can't even afford to buy you and the boys food. How could I possibly think about asking you to marry me now?"

She listened to me, was quiet for a moment, then said, "Well, are you going to marry me or not?"

I tried one more time to get some reason into this conversation, "It's hard living in Wyoming. It's not like Oklahoma."

"Make up your mind. Is it yes, or is it no?"

"Yes, I will marry you. You do understand that as long as I am going to college I will have nothing? It will be difficult to provide for you."

"Yes, I understand perfectly."

A couple of days later, I received a letter from Teresa. She wanted to meet my parents before she made her final decision to marry me. It was important to her that my mother liked her and her boys. She had experienced a bad mother-in-law, and she wasn't going to do it again. She had to be sure that she and her boys would find a loving family.

Teresa told me that if everything worked she would sell her home and we could live off the profit until I finished college.

Just like that, the process of courting was over for Teresa and me. There was going to be an end to phone sex. Finally.

When I recovered from the shock of what I had done, I called Teresa.

"What about I bring my parents down to Oklahoma at Easter time? I think they have a trip planned to go to Texas to visit my aunt and uncle. Lawton is on the way."

It was arranged. Teresa would meet Mom and Dad on Easter in Lawton, Oklahoma, under her roof and on her turf.

So we showed up at Teresa's home just before Easter. I knew Mom and Dad would like Teresa and her children. I was pretty sure Teresa and her kids would like Mom and Dad, too. But I was nervous.

Teresa had spent days preparing for our late-afternoon arrival. She had enlisted the help of her friends to make sure everything was perfect. Supper was ready when we walked through the front door. I could tell right off that it was good, all around.

Teresa opened her red front door and Matthew and Jeremy rushed out to greet me and my parents. Perfect.

As soon as the introductions were over, Teresa invited us to sit down for dinner. The table was made up with candlelight and fine china. Teresa placed me at the head of the table, she took the other end of the table and placed Mom on her right and Dad on her left. Matthew was seated between Dad and me, and Jeremy would be between Mom and me.

The adults had just sat down when Jeremy came running into the dining room and said: "Mommy, there is water on the floor in the bathroom."

Dad and I jumped up and ran for the bathroom. The toilet was overflowing and the bathroom floor was flooded. I ran to the toilet and shut off the water valve. Dad turned to Teresa and asked her if she had a shovel. She said she did and ran to get it. The shovel was a scope shovel.

Dad took the scope shovel and started shoveling water into the bathtub.

Teresa stood there dumbfounded and devastated. She had worked so hard to make a good impression on her prospective parents-in-law. Now her prospective father-in-law was shoveling sewer water into her bathtub.

Mom took charge of Teresa. She had her help gather all the spare towels in the house, and the two of them built a dam to stop the water from reaching the living room carpet.

Teresa said, "I have a plumber; I'll call him."

Dad responded, "No need. We'll fix your toilet."

It ended up that Dad and I had to remove the toilet, take it out into the front yard, turn it upside down, and flush it out with a garden hose. The neighbors watched with interest. When we found a toothbrush stuck in the toilet, we guessed that either Matthew or Jeremy had been the flusher. After we got the toothbrush out of the toilet, we went to the lumber yard, purchased a new wax ring, and reinstalled the toilet.

Mom and Teresa got to know each other better by cleaning up the bathroom together. Dad and I were proud to announce the toilet was good to go.

An hour or so later I flushed the toilet. It worked perfectly. I said, "There you go, my dear. We are back in business."

We sat back down to eat a cold, overcooked supper.

I said something like, "So, Dad, Mom, what do you think of Lawton?"

Dad looked at me and then at Teresa. "I like it just fine," he said. Mom winked at Teresa.

Teresa smiled at me. With her eyes, she told me she approved. And my dad and mom responded in kind. They were no fools; they had just met a beautiful and gracious young woman and two beautiful soon-to-be grandsons.

My stress evaporated.

Dad took me aside later that evening and said, "You'll have to deal with me if you ever do anything to hurt her."

"Dad, you don't have to worry about that." I knew he meant what he said.

The next day, Mom and Dad left for Texas, and the only official difference between me and a married me was a preacher.

But something else was going on that weekend. Something that brought dark clouds to my world. The last of the American evacuation of South Vietnam was in full force on Easter Sunday, 1975. The TV news was hard to watch.

The hurt I felt was deep. It made me realize how wrong I had been during my discussions with Captain Thinh back in 1973.

I recalled the conversation I'd had in 1970 with Sergeant Son, his dad, and his grandfather. With them and with Thinh, I'd been so wrong. With the last chopper out of Vietnam went the essence of America's honor. My belief in the goodness of our way of life had been shattered.

After Dad and Mom left, we played with the boys and watched TV until it was time to go to bed. The news on TV upset me, and Teresa sensed my stress. She held me when we went to bed. She understood that I hurt real bad. I couldn't tell her why, because I didn't know. Or maybe I did and was too ashamed to admit it.

I had waited five years to have someone hold me while I hurt. There were so many reasons to hurt. It was a great comfort to feel her holding me. For the first time since 1970, I cried.

—

In addition to being ashamed, I was mad.

There were valid and logical arguments, pro and con, for our Vietnam decisions and actions in the 1950's, the 1960's and the early 1970's. But to my way of thinking, there was no just argument in favor of what we had done to our friends—and to ourselves—in Southeast Asia in the spring of 1975.

I believe that of all the mistakes America has made in the last hundred years, none were as devious as what we did to South Vietnam. Nothing else has shown more clearly our failure as a civilized nation than the day we sold out Son's and Thinh's beloved nation.

In 1973, Thinh left America to return home and continue the fight. I never forgot the long discussions we had about America. In 1970, Grandfather Son and Mr. Son had told me there was something fundamentally wrong in America. On Easter 1975, I realized I had received the wise teachings of several very astute men. They could see the world, my world, without my bias. They were right and I was wrong.

For forty years I have asked myself why I couldn't see what was coming. I have no answers. For forty years I have felt guilty about what I said to Thinh and Son, and to Son's family. I was not there for them in their hour of need.

And now they are gone and it is too late to say that I am sorry. I may never be able to close this loop.

Cu Chi, South Vietnam
February 1970

The morning was still young when I met my first ghost. He would haunt me the rest of my life.

The sun was just jumping up over the eastern skyline. We left battalion in my Jeep and headed for Tay Ninh. It was just my driver and me. This was to be my first trip into a hot combat zone, but still, I had no reason to suspect my life was about to change. I was thinking about snipers and roadside bombs, and how vulnerable I was in the Jeep.

In Cu Chi, we noticed a crowd in the street. I told my driver to stop to see what was happening.

The body of a young man lay on top of a pile of corpses. A few hours before, he had been strong and tall. He looked like he had been healthy. Probably a good soldier.

Looking a little closer, I could tell he'd been gut shot. I remembered being told that gut shots caused painful deaths. To look at him, you wouldn't have thought it hurt all that bad. He looked hopeful and serene, like he was having a pipe dream. When the opium wore off, he'd expect to be waking up to a better, brighter day where he'd be the one looking at piles of American corpses.

And then, I swear, he turned his head and spoke to me.

Listen, my friend, he said. *I know you haven't been here very long, but it is time for you to wake up. Things are not what they seem. If you don't get your shit together, this could be you.*

Things were going too fast.

In two years, I'd been a bachelor officer in Germany, then a family man with a few months left of active duty, then a Captain headed off to Vietnam. I'd accepted the promotion so I could take better care of Twinky and Anna, even though I knew it meant I'd end up in Vietnam.

Now here I was, standing in the street in Cu Chi, looking at a pile of dead humans and confronting my first ghost.

I'd once been such a fun-loving guy. The transition hadn't taken very long.

It was common for the Army of the Republic of Vietnam (ARVN) to display the bodies they accumulated the night before in a town or village square, then wait to see who reacted. It was a way to identify enemies.

On top of the pile was this kid who appeared to be an example of a fine soldier. Now he was dead. At least I'd thought so before he started talking to me.

Listen, my friend. You are at war. They mean business. There are advantages to being an American in Vietnam. But there are some disadvantages. There are piles of dead bodies, and there are piles of dead bodies. You know there are piles of American soldiers somewhere out there, too. Take care you don't end up on one.

I was stunned. Not only was this my first pile of dead bodies, it was my first experience with a world I did not want to acknowledge. I was just a little pissed how brash that asshole dead NVA (North Vietnam Army) soldier had been to me, a captain in the U. S. Army.

But as the cocky, young, and patriotic officer I was, I knew I could handle anything that came down the pike. In part of my mind, there was no real doubt that I could handle it, no matter what it was. But in another part of my mind, there was a little confusion.

I found it unsettling.

It was strange that I could only focus in on one body in a pile of corpses four feet high. His eyes had almost closed now, but I could still feel him looking at me. What was he trying to tell me?

"We are not finished, you and I," he said. *"Catch you later, okay?"*

Wyoming, June 1975
Peace in my little war becomes possible.

On a beautiful, sunny spring day, at an outdoor chapel in a valley full of spring flowers and deep, green grass in the Medicine Bow Mountains of southern Wyoming, we stood in a small grove of aspen trees. As a nearby river moved slowly past us, a preacher looked at us over an altar made of hewn logs and asked Teresa, Matthew and Jeremy to say: "I do."

They said: "I do."

Finally, I had a home.

Phillipsburg, Germany
1967–1969

Prior to receiving orders for Vietnam, I had been a Special Weapons Liaison Officer with the West German Army. Since December of 1967, I had been assigned to NATO as a team commander in a small US Army detachment. We were co-located with our supported divisional artillery of a German army division. We were stationed in Phillipsburg, a little town on the Rhine River in Central West Germany. Our job was to secure and maintain assigned nuclear artillery weapons for our NATO (North Atlantic Treaty Organization) partners. If the Soviets and their allies attacked West Germany, it would be our jobs to support a West German army division and supply them with nuclear weapons.

Because of the sensitivity of the mission—from a tactical point of view, a security point of view, and a political point of view—the troops assigned to our detachment were, for the most part, good at their jobs. Our detachment was located not too far south of U.S. Fifth Army Headquarters located in Frankfurt. The detachment was physically located on the east bank of the Rhine River about halfway between Heidelberg and Karlsruhe. We were easily accessible for visits and inspections from the Department of

Defense, Joint Chiefs of Staff, congressmen and other politicians who had an interest and the need to know about the security of the nuclear arsenal allotted to NATO. We got a lot of visitors of one kind or another so it was to everyone's advantage that we looked good most of the time. The Germans had a vested interest in us being good at what we did. The Germans we worked with were good at what they did, too.

We got a view of Germany that few American military types would ever see. We lived and worked on what we called "the economy." Most American military worked and lived in American compounds. The American compounds were self-contained. Everything the soldier needed to work and exist was contained in the compound. They had little contact with Germans on a continual basis. I had a cousin who had been in Germany for two years. He and his wife lived their life on the American compound in Karlsruhe. I visited them in January 1968 and suggested we go into Karlsruhe to eat dinner. At first they did not want to do it. The American army built cities within cities in Germany. They were called kasernes. These kasernes provided for all of the troops' living needs during their stay. They didn't have to live or shop on the economy, meaning they didn't have to go into greater Germany. There seemed to be a phobia of the unknown, of trying to communicate with someone who didn't speak English. It was easier to stay in the kasernes then to explore this foreign culture. I had to almost beg them to go on this adventure with me.

I was surprised to learn they had never eaten in a restaurant on the economy. They could not read a German menu and they could not speak German. I had been in Germany a little over one month and I had to order their dinner for them. They were surprised at how good the food was.

Many American soldiers had little or no contact with German soldiers during their two-year tours of duty. American soldiers who did not have contact with German soldiers tended to consider them as second class. Their attitude bordered on hostile.

I found my German counterparts to be really good soldiers.

The Germans had a word for American soldiers and Americans in general. It dated back to the Second World War. The word was "Ami." The word was not intended to be flattery.

We even began to see our own countrymen as the ugly Americans we had heard of so often. Americans were loud, rude, disrespectful and inappropriately dressed. This opinion extended to soldiers and to tourists. Tourists tended to be long-haired, smelly college students hitchhiking across Europe during the summer. The German stereotype of the American hippy. We didn't see other Americans much, and when we did, we didn't like much of what we saw. I ran into some "Ami" on occasion.

I had been assigned to this duty station directly out of the Field Artillery Officer's Basic Course at Ft. Sill, Oklahoma. When I arrived in Germany, I was a snot-nosed "butter bar"—military slang for Second Lieutenant, due to the gold bar insignia on the uniform. I had no experience behind me other than ROTC and the Basic Course. I didn't know my ass from a hole in the ground. I was twenty-three.

Luckily, I was assigned to a team with a good non-commissioned officer (NCO). He guided me through those first months until I was able to stand up and take control. He did a good job.

The liaison assignment was like a dream. I was working with good NCOs and good enlisted men. We all knew exactly why we were there and what we were there to do. We were a small detachment of American soldiers located deep within the interior workings of a German combat division. We provided the German division with significant combat fire power that we all hoped we would never use. We were there because the Germans wanted us, needed us there for protection against the Soviets. The Germans treated us well. I began to think that Army life was something I could get used to.

There were other benefits.

Take Fasching, for example. Fasching is kind of like a carnival for religious folks. It is a progressive party that starts in November and goes through Easter. It is a come one, come all kind of affair that has room for the believers and non-believers, Saints and sinners, Germans and Americans.

Phillipsburgers liked their Fasching.

My first experience with Fasching was in January of 1968. The festival was in full swing. I was in the balcony of the Phillipsburg Fest Hall one Friday night with a few of our detachment soldiers and a couple of German soldiers we supported. I was looking out over the dance floor, drinking a local sparkling wine (we called it Sparkle) and watching everyone dance to an Oom-pah band. Suddenly this pretty young thing pushed her way in front of me, put her arms on the banister, and looked down at the dance floor.

I took a step back, thinking that was a little rude. I stood there looking at her back for a minute, not knowing what to do next. Then she turned around and spoke to me in German.

One of the soldiers with me said, "Lieutenant, she is asking you to dance."

I smiled at her and said okay.

Down on the dance floor, we'd jump around until the band stopped playing in the middle of a song. Then we'd put our hands in the air, yell, "WOOOOO" and fall to the floor. We'd lie there until the band started playing again, then we'd get back up and resume jumping around. I had no idea what we were doing or why we were doing it, but, we were having a lot of fun.

After a while we got tired, so she led me back to the balcony and to my little group of intoxicated idiots. She said something in German, turned her back to me and walked out of my party. I watched her tight little butt sway as she stalked away.

I stood there wondering what the hell had I done wrong. Then I became aware of some laughing behind me. I turned around.

My little band of merry idiots were slapping at each other and laughing at me.

"What's up?" I said.

"Lieutenant, you just met 'The Ghost'," SGT Pollard said. "She likes you. You like her?"

"She's kind of cool, why?"

SPC Rains said: "She's a spy." He started laughing again.

Pollard said: "The only time she comes around is when we have a bachelor officer at the detachment. That's why we call her 'The Ghost'." He looked at Rains, winked, and offered me another glass of Sparkle.

Well, fuck, I thought. I shook my head and started laughing with them.

The standard operating procedure required us to report suspicious activity to the detachment commander, Captain Noel, so first thing Monday morning I went to his office to tell him I'd been in contact with a spy. When I got to the part about her nickname, he nodded.

"Oh, you mean Secklinda," he said. "Yeah, we know about her. What do you think?"

"She's kind of foxy."

"Anything else I need to know?"

"No, sir."

"Well I suggest you enjoy the ride. Just remember to be careful what you say around her." He smiled and dismissed me.

As I walked out of Captain Noel's office I smiled and thought, *Well hell, if I am going to get my horns trimmed, I might as well get 'em trimmed by an expert.*

Over the next several months, as she would launch my satellite, I would think, *Keep your damn mouth shut.*

At some point I decided I needed to test Secklinda's loyalties, find out what kind of spy she really was. I wanted to know how much she knew about what we did in Phillipsburg.

Our relationship was kind of complicated. She would never let me take her home, using the excuse that her parents hated Americans, so we always met in a public place. I never learned exactly where she lived and I never learned where any of her friends lived. Her refusal to let me into her personal life just whetted my appetite to find out more about her.

On this Friday night we planned to go to Heidelberg to party. I picked Secklinda up in my car, a 1967 Plymouth 383 four-barrel that would haul ass down the autobahn with the best of them. She liked riding in my big American car.

I said to her, "Before we leave for Heidelberg, I need to check the guards on duty at the storage bunkers."

Everyone in Phillipsburg knew about the storage bunkers, and they thought they knew what was stored inside. They knew about Hiroshima and Nagasaki. Many of them had firsthand experience with nerve gas and blister agents in the First World War, and they knew about anthrax. They knew the bunkers were full of that stuff. They treated the area as if it was contaminated with some incurable disease.

In a way, it was.

No Phillipsburger would get anywhere close to those bunkers. The bunkers were off limits to all but our small detachment of Americans and our German army counterparts.

I wanted see how The Ghost would react when I took her out to the bunkers and left her alone in my car, parked just outside the wire.

She about shit.

I guess it wasn't very nice, what I did to her.

Emergency Demolition Training 101
Federal Republic of West Germany
Phillipsburg, Germany
Spring 1968

As my drill instructor called me just after I was commissioned, Brand-new, Wet-behind-the-ears, New-to-the-world, Snot-nosed, Butter bar, Second Lieutenant, One-each, Martin reported for duty at the 3rd USA Missile Detachment in Phillipsburg, Germany, December 1967. It was a good description of me.

The 3rd U.S.A. Missile Detachment in Phillipsburg was assigned to SASCOM—Special Ammunition Support Command. SASCOM provided nuclear weapons support to members of NATO. This translated to Butter Bar Martin providing liaison support to the Army of the Federal Republic of West Germany.

Or here's another way to put it: our little detachment of thirty-some soldiers looked after several eight-inch nuclear rounds and several Honest John rocket nuclear warheads for a German Army division that was to help defend the Fulda Gap in case the USSR and the Warsaw Pack decided to attack.

Our thirty-man detachment was co-located with units of two German artillery battalions and a German infantry company. Our peacetime job was to train the artillery battalions how to move and shoot eight-inch nuclear rounds and to move and shoot nuclear-tipped Honest John rockets. We were charged with the mission of storing and securing the nuclear weapons right in the middle of two thousand German soldiers.

Pretty heady stuff for a brand-new second lieutenant. Even more so given the fact that most of my classmates were headed to Vietnam, where they were given a shelf life of seven minutes in direct combat.

As commander of the 8-inch artillery assembly team, I'd been assigned the shit duties in the detachment. Since I was one of only three officers in the detachment, I had several of these shit duties. One of them was that of emergency demolition officer.

It was my duty to store and maintain all the emergency demolition (ED) materials we'd use if a situation put us at risk of losing our nuclear weapons. I had to make sure we had all the C4 plastic explosives, 40-pound shaped charges, 15-pound shaped charges, detonation cord—we called it 'det cord'—ignition fuses, and the hand tools we needed to assemble the demolitions to blow up all the nuclear stuff. I had to train the thirty or so men under me on how to set up a ring main of det cord to each nuclear weapon and its associated components so that one blast would destroy everything.

Again, I had to do all this while two thousand Germans stood around and watched.

So there I was, with close to 2500 pounds of really cool firecrackers, about to conduct my first pre-inspection inventory. I counted the shaped charges; they were all there. So were the boxes of C4 and the fuse and igniters.

All good, except for one detail I'd missed before. Everything had been manufactured in 1944. The same year I was born.

Maybe that was why the C4 was flaking off inside its plastic wrapping material.

Hell, I thought. That didn't seem right, but what did I know? I checked the manual. Not a word about age or flaky C4. So I called Battalion for guidance. They said they'd get back to me.

Turned out nobody knew if the stuff was any good or not. Keep in mind that while I was still pretty damp behind the ears, I was talking to a lieutenant at Battalion who was at least somewhat dryer behind the ears. But he'd never used this stuff before, either. Or even seen it, as far as I knew.

Here's how Battalion solved it. He authorized me to order a new complete inventory of ED materials to replace the present inventory. I submitted the order. Battalion approved it. A few weeks later, a whole new set of toys arrived, and I headed to the bunker to inventory them as they were delivered.

Everything was cool. At least I thought it was until I saw it.

The new ED materials turned out to be two years older than the first batch. These had been manufactured in 1942. The C4 was just as flaky. I had a double allocation of flaky stuff that may or may not have worked.

So I call Battalion again, and we butter bars discussed what to do next. I had way too much of this stuff and I needed to get rid of the extra. My recommendation was to ship it back to the depot. 2nd Lieutenant Robertson informed me that would not happen. He instructed me that since we had access to the German demolition training range, I was to use the surplus ED materials for training.

That I could do. *Fourth of July, here we come.* I put ED training on the schedule, contacted the Germans for permission to use the training range, and asked them for a box of blasting caps. They stored them for us since we couldn't keep the caps with the other stuff, for obvious reasons. I always wondered how we'd get blasting caps if the Germans decided to take our nukes. I guess it must have made sense to somebody.

I ran into one slight problem. The maximum amount of explosives the Germans would allow for any one blast was three kilos. I had 2,500 pounds of explosives to use up. At three kilos a shot, I'd be retired from the Army before I'd get through all that.

Three kilos wasn't realistic in terms of what we needed for training purposes, anyway. And on top of all that, nobody in SASCOM had ever set up and detonated a realistic ring main to see if the concept would actually work.

At least, not as far as I knew.

In the spring of 1968, I was of the opinion that I was just the guy to do it. I'd pretend I didn't know about the three-kilo rule. Robertson had given me full authority to blow up 2,500 pounds of explosives, so I'd do it the way the manual told me to.

Training day came in late June, as I remember. It was a beautiful day, temperature in the eighties, no wind, everything green. Everything was perfect. Best of all, no one knew what I was up to except those who were directly involved in the training. Like me, they couldn't wait to get started.

Step one was to run a thousand meters of det cord in a big circle to establish the ring main. Then we ran several feeder lines of det cord to some old car bodies in the demolition area. Once the cord was set, we placed a box of C4 in the trunk of one car and a 40-pound shaped charge on the engine of another. We set a 15-pound and a 40-pound shaped charge next to each other on the ground to compare the depths of the holes they made, and a box of C4 for the same reason. So on and so on until all 2,500 pounds of my firecrackers were in place.

I told my fuse guy I wanted ten minutes of fuse for a dual ignition system. This meant I'd put a blasting cap and an ignition fuse on each end of the ring main and ignite two fuses each for ten minutes of burning time.

While my fuse guy was figuring out how long the fuse needed to be to burn for ten minutes, I checked the ring main with Staff

Sargent Ellison. Everything seemed to be connected the way the manual required it to be. I was about to pronounce us ready to go when I noticed a shepherd about two hundred meters away, grazing his small flock of sheep on the side of a hill that overlooked our position.

As far as I knew, SSG Ellison was the only one there who spoke any German. It was a joke in the detachment that his German was limited to *Habenze ein ashenbecker, bitte?* Which meant, I think, "Do you have an ash tray, please?" Ellison was a frequent guest at the local hofbrau haus, he drank a lot and he smoked a lot. Ashtrays were important to him. Not much else.

Anyway, I instructed him to go over to the shepherd and tell him there was about to be a big boom. I watched as Ellison walked up the hill, shook hands with the shepherd and carried on what looked like a conversation. When they were finished, they shook hands again and Ellison started back down the hill. The shepherd watched him walk away and he looked at me and waved. I waved back. When Ellison got back he said the shepherd didn't seem to be worried about it. His sheep were used to three kilos worth of boom, Ellison told me.

Okay, I thought. *He's been warned.*

My fuse man, Private First Class Hill, came up and told me he'd rigged the dual-ignition system for ten minutes of burn time. Now, you need to keep in mind that he'd never done this before, and neither had I. I had no idea how long a fuse needed to be to burn ten minutes. I assumed he did. But I was wrong.

I ordered everyone to evacuate up the hill and stand under this big tree to watch the detonation. When they'd removed to what I considered a safe distance—I had no idea what a safe distance really was—I looked around and noticed the shepherd was sitting on a tree stump watching me from the other hill.

Surprise, surprise, I thought. *Have a nice day.*

Everything was ready.

"Fire in the hole!" I yelled. I repeated it twice more, and then I pulled the pins on the igniters. The powder in the igniters burst, catching the fuse on fire. The fuse began to burn down like the fuse on a firecracker, only much bigger

Damn, I thought as I walked up to the tree to join the others. *This is fun.*

When I reached the tree I noticed Captain Cintron, our new detachment commander, standing among the men. He was a Puerto Rican who had just returned from Vietnam. I mention this to set the stage. CPT Cintron was a little high strung. You might even say hyper.

As I approached him, he asked me, "How long before it goes off?"

I looked at my watch. "About five minutes and thirty seconds, Sir."

CPT Cintron looked down the hill. He pointed at the smoke from the burning fuses. Then he walked around the tree and unzipped his pants to take a leak.

Thirty seconds later, a thousand meters of land disappeared under this big ball of flame, smoke and dust. The rear end of a car came hurtling out of the cloud and crashed to the ground about a hundred meters away.

That was close, I thought.

What I didn't think about was that there would be a shock wave. There was a big one. It came up the hill, hit the tree we were standing under, and passed on by. Dust and leaves began filtering out of the tree, settling on our uniforms and turning us all a pale shade of tan.

For a few moments everyone stood transfixed at the spectacle before us. Then I remembered CPT Cintron. Since he was my commanding officer, I felt it necessary to get his feedback as quickly as possible. When He was on his knees with a hand against the tree trunk, in the process of getting up. His uniform was covered in piss and mud. He caught my eye, frowned at me, turned and walked away without a word.

Then everyone woke up and started jumping and yelling and having a great time. It was the best training they'd ever had.

I thought, with a little reservation, that maybe I used too much explosive. Then I grinned. *I hope he isn't too pissed off,* I thought.

SSG Ellis walked up to me, smiling, and said, "Sir, you got to know that was impressive. Look at those sheep hauling ass."

I looked over to the other hill and all I could see was the back of the sheepherder running after his sheep as they all un-assed the area.

We both laughed.

I said; "Sergeant, I'm up for promotion in a couple of months, what do you think my chances are after today?"

SSG Ellis said; "Sir, you ever consider what life would be like as a Private, E-1?"

We laughed again.

My immediate after-action report to my commanding officer, CPT Cintron, would be something like this: "Sir, we are trained in ED procedures, and that old stuff is just fine."

PFC Hill was still dancing around. I walked over and said, "I thought I told you I wanted a ten-minute, dual-ignition fuse system."

"Yes, Sir, you did. So I figured out how long a ten-minute fuse needs to be and cut it in half. Five minutes plus five minutes equals ten minutes."

Now, how could I argue with that logic? I smiled at Hill. He smiled back and went on celebrating.

We waited about 30 minutes before going into the blast area, just to be sure there would be no delayed explosions. Some of the car bodies were nothing but bent frames, but a few of the car bodies were still identifiable. The damage caused by the shaped charges caught most of our attention. The purpose of the shaped charge in our ED procedures was to melt a hole in the side of the Honest John nuclear warhead and blow the nuclear core out the other side without causing a nuclear detonation. To test this theory, we'd placed a shaped charge on a car engine. We found that the shaped charge did exactly what we were told it would do. The front part of the engine was missing, part of the remaining engine block was melted, but the rest of the engine was not damaged.

To test the power of C4 plastic explosive, we'd put one flaky stick in the front seat of a car, two flaky sticks on a tree stump, and a box of C4 in the trunk of a car. The car with the box in the trunk was no longer evident, and the other two cars were no longer salvageable.

The thousand meters of det cord had left a nice clear path on the ground, completing its work faster than Superman's speeding bullet.

All in all, we were very impressed with our work that day.

I got back to the detachment about two hours after the blast. I learned the shock wave had hit the brand-new communications center building and cracked a wall. It had also caused all the open windows to slam shut, but luckily no glass was broken.

Not so lucky for the brand-new office building that had just gone up at the Goodyear tire factory about a mile away. The blast had shattered all the glass windows on the side facing us.

When I learned that the glass was gone from the Goodyear office building, I felt a little guilty. For about thirty seconds. Thank God I had 2LT Richardson's authority to cover my ass. I hoped nobody would ask me if I knew about the three-kilo rule.

Lucky for me, nobody thought to ask.

The next morning, CPT Cintron called me into his office and told me that the people at Goodyear were upset. He considered it appropriate for me to apologize for breaking all their windows. I didn't see it that way. I had authority from Battalion, from 2LT Robertson, to do exactly what I'd done. I didn't see why I needed to apologize to anyone.

He did not tell me to clean his uniform, so I figured that issue was closed.

Technically I was right. But from a good German/American relations point of view, I was wrong. Someone was going to apologize. I never learned who got the honors. It could have been CPT Cintron, or the battalion commander, or the group commander.

All these years later, I still think that day was a hell of a lot of fun.

It took about a week for things to get back to normal in the detachment. I noticed that there was no shortage of volunteers for the projects I had to complete. At least from the detachment soldiers' point of view, the wet-behind-the-ears part of my unofficial title had dried up. I was now just a butter bar. When I came up with a new training idea, they were ready to volunteer.

However, I did not get off without suffering reprisals for causing some political issues within the battalion. Someone picked up on what appeared to be my lack of sensitivity to the job of representing the United States Army in a foreign land. They couldn't shoot me for being stupid, but they could and did make my life miserable.

The worst assignment an officer could draw in our battalion in 1968 was that of defense counsel in a special court martial.

There were three levels of military justice. The first level was called non-judicial punishment. This was used when someone did something stupid, like getting too drunk. It was a slap on the wrist-type punishment. The second level was a special court martial, for when someone did something stupid like get drunk and (without cause) kick the crap out of a German civilian. This level of punishment could result in some time in the stockade and a less-than-honorable discharge. The final level was a general court martial. This was for when someone did something stupid like commit murder or desertion. This level of punishment could result in the firing squad.

Because of the minor nature of the acts and punishment involved in the first two levels, military discipline was handled within the battalion using battalion assets. But the third level was serious stuff. It required officers who were real lawyers from the Judge Advocate Section found at higher commands.

Within the battalion, the preferred method of military discipline was Article 15, *Non-Judicial Punishment.* The detachment commander could fix the problem without getting other battalion assets involved. The second level of discipline required several battalion officers to take time out of their regular duties to prepare for and conduct the court martial. The battalion commander made the decision whether or not to convene the court martial.

The battalion commander was not in the habit of convening a court martial when there was a probability that the decision would be not guilty. A not guilty decision meant the court martial was probably unnecessary, a waste of time and resources. There was no upside for the battalion for a not guilty verdict.

So there had to be overwhelming evidence of guilt for the battalion commander to convene a court martial. We all knew that the assignment as defense counsel in a court martial was a no-win assignment. If you won a not guilty verdict, you made the battalion commander look bad. If you got a guilty verdict for the soldier you represented, the soldier would go to the stockade and you would look as stupid as everyone thought you were.

I got assigned as defense counsel a couple of times.

All I knew about the Uniform Code of Military Justice (UCMJ) was that there were a bunch of do's and don'ts contained in a big, thick book. I figured it was my duty to read parts of the UCMJ book. And I tried, because I knew some of the don'ts could get someone hung. Normally when I started reading that book I would be asleep within fifteen minutes. I did read enough to learn that it was not good to be a defendant in any court martial.

There was a standing joke in the court martial business that went something like this: The trial judge would open the trial and say, "Bring the guilty bastard in—oops, I mean bring the defendant in." In reality he might think it, but since the battalion commander was his boss, he would never say it.

As the guilty bastard's defense counsel, I had to meet with him to develop a defense strategy. They housed the defendants in an American military police prison stockade facility. Now, the Germans knew how to build prisons. Good prisons. Dachau, a Nazi facility dating back to the 1930's, was one of those good German prisons, and it was the perfect choice for the American military stockade. So in 1968, I found myself going to Dachau to interview a soldier so I could prepare to defend him in a court martial.

Before I signed into the stockade to interview my assigned defendant, I decided to visit the Jewish memorial at Dachau. There was a big metal sculpture of a large number of dead persons or dead departing souls. As I stood there looking at it, I felt I was being drawn into the sculpture, maybe becoming a part of it. I took some time to visit the memorial museum. It was not an enjoyable experience and it made me nervous. I found myself wishing I had not visited the memorial before I had to go inside the stockade.

After I finished with the museum, I signed into the M.P. station. I was given an escort.

As I was being admitted into the stockade compound, I felt a little uncomfortable. I was tense and ill at ease.

I noticed grooves in the concrete where so many humans had stood and moved that the surface of the concrete was worn away. I was escorted into one of the concrete barracks-type structures and placed in an interrogation room just big enough for a small table and two chairs.

My escort said he would bring to prisoner to me. Then he closed the door and I was alone.

The skin on my back began to tingle. How many dead people had been in that room? I felt as if they were still there. I wanted out of that place.

When the door finally opened and I felt some fresh air surge into the little room, I took a deep breath. My escort opened the door wide enough for my defendant to enter the room. My escort placed the defendant in the other chair and chained one hand to a metal ring on the table.

I introduced myself and explained why I wanted to interview him. I asked him if he was willing to talk to me.

"Yes, Sir."

"Have you read the charges in the court martial?"

"Yes, Sir."

"Do you deny the charges?"

"No, Sir."

"Are you telling me you want to plead guilty to the charges?"

"Sir, what I am telling you is I want out of the Army and I'm willing to do whatever it takes to get discharged."

"If you get discharged because of this, it will not be an honorable discharge."

"Sir, I don't care."

"You will have to live the rest of your life with a dishonorable discharge on your record. Are you willing to go that far?"

"Yes, Sir."

"I recommend that you try to fight this."

"No way, Sir. I will not risk spending any more time in the Army than I absolutely have to."

"You will probably get some stockade time."

"I will risk it. All I want you to do is tell the powers to be that I want out of the Army as quickly as possible."

"Are you sure about this?"

"Yes, Sir."

"Okay, do you have anything else you want to tell me?"

"No, Sir."

"This interview is over. The trial is next Wednesday. I will talk to you before the trial to make sure you haven't changed your mind."

I stood and walked to the door and knocked on it. My escort opened a little window in the door and I told him we were through with the interview. The door opened and a guard took my defendant out of the room. My escort took me out of the stockade back along the grooved concrete, processed me out, and left me alone on the free side of the stockade.

I stood there looking around, breathing fresh air, feeling free and very upset. I was the instrument that would destroy a young man's future. I didn't like that feeling.

—

As a young officer just starting to think of a career in the Army, little things began to bother me. I volunteered expecting to go to Vietnam. Instead, I was sent to Germany to live the easy life of a liaison officer. I was ashamed that the only ribbon I had on my uniform was the National Defense Ribbon. Everyone had one of those. I did not have a Unit Patch on my right shoulder that announced to the world that I was a blooded soldier. I was not a combat-qualified officer. I was a good-times soldier who would spend two years in the Army and never fire a weapon in a hostile environment. I would never know what it was like executing the power of life and death or what it felt like to stare death down. As an officer in the United States Army, I felt incomplete.

I was not a member of the elite. I wanted very badly to be a member of that secret society I'd witnessed as a boy.

Being assigned to a German garrison had other effects. To a large extent we were cut off from the rest of America. There were no American TV news broadcasts to watch, and no American radio stations except for Armed Forces Radio.

You haven't lived until you watch *High Noon* in German. It's a hoot!

There were no American newspapers, except for the *Stars and Stripes*.

Aside from the very large events in America, like the murders of Robert Kennedy and Martin Luther King, the race riots, the landing on the moon, the Tet Offensive in Vietnam and the body counts on both sides, we heard very little about what was happening outside our little world in Phillipsburg. What we did hear was heavily censored by the Army or biased to the European point of view. Our German friends' lack of background and understanding of America was staggering; we found their viewpoints suspect. So for two critical years, I was not in tune with what was happening to my country or my Army.

As a liaison detachment living and working with the Bundesweir (West German Army), we formed close working relationships with our German counterparts. We figured we were pretty damn good, and so were the Germans we worked with.

We officers—there were three of us, one captain and two lieutenants—were often invited to the German Officer's Club for drinking and target practice. The Germans loved getting drunk and getting the guns out to see who could shoot best. Due to duty assignments, only two officers were ever able to go at the same time, and sometimes it would be only one.

The club was in the basement of one of the German barracks, in the same room they used as the barracks' bomb shelter and fallout shelter. It was constructed of thick, reinforced concrete walls, ceiling and floor. The booze never varied for us. We always got a shot of Stienhager and a boot of beer. The Stienhager tasted like a good grade of gasoline and the beer was Phillipsburger Beer, direct from the Rhine, the most polluted river in Europe. Just one smell of the Stienhager and I'd get dizzy.

The Germans loved getting us drunk, which was not very hard to do. When we were appropriately liberated of our senses, they would get their pistols out to test our shooting skills. Our German officer counterparts figured they were better shots than us. They'd place some sort of a target, a picture or a plate or something like that, at a concrete wall at the other end of the room and we would stand next to the bar and start blasting away.

We shot up the Officer's Club more than once. They were better shots than me.

Generally after the fun and games, it became time for the Germans to start an argument about America. It was usually them against us. We would argue about our role in Vietnam. We took the position that we were trying to do the same for South Vietnam as we did for West Germany. They took the position that what we were doing was immoral and was hurting our image in the world. I found it a little strange hearing this argument from Germans.

During my last visit to the Officer's Club, we finally got into the real meat of their dislike for America—the miserable way we treated blacks. They thought it was terrible, that we were extremely racist. Our position was to ask them to give us an example of racism that they saw in our Army. They could not. But they were not ready to admit that blacks held all the ranks in our Army and that there was a close relationship between the whites and blacks in our detachment. I told them we knew we had some problems and that we were trying to correct them.

That evening, one of the German firing battery commanders who we seldom saw was at the club and he joined the argument. We Americans looked on him with real suspicion because he was a member of the Neo-Nazi Party. He carried himself in that stereotypical Nazi style that left us with the opinion that he felt everyone in the room was inferior to him; that the Second World War was not over, it was just postponed.

He didn't like us, either.

The Stienhager flowed, the conversations got louder and louder, and the tempers got raw. I began to warm to the challenge of defending America.

Finally, the Neo-Nazi asshole spoke directly to me. He called me racist.

"Listen, asshole," I said. "We have a problem, I'll admit that. We all have a skeleton in our closet. At least we're trying to bury ours peacefully. We're not trying to exterminate it."

Then, for emphasis, I leaned forward, getting my face as close to his as possible, and said, "I have been to Dachau. Have you?"

Nobody said anything. Except for the music coming from the jukebox on the other side of the room the room was silent. I looked from face to face, all but one registered surprise and disbelief. One face registered pure hate. His hands were gripping the edge of the table in front of him and his knuckles were white. His jaw was flexing but his mouth remained closed.

    I took a sip of beer, set the glass down on the table, stood up and excused myself.

Laramie, Wyoming
February 1976

On a clear, cold day with several inches of snow on the ground, I was driving home after picking Teresa up from her job at the Bureau of Mines at the University of Wyoming.

I was concentrating on the task of driving when Teresa said, "I'm pregnant."

Back when we were about to get married, I'd told her two boys were enough. We did not need any more kids.

She'd said, "Yes, dear."

I assumed, since I was man of the house, my word was final. That was that, I thought. End of discussion. Problem was, we'd had that conversation before the preacher talked to us. Before the preacher, my word was final. And it was final until I said, "I do."

Then my word, without my knowledge or approval, became negotiable.

Sometime in late August or early September, after an hour or two of body painting and other stuff, Teresa casually mentioned that she had been on the pill for a long time. It was her opinion that she needed to give her body a rest. There were other forms of birth control that worked just fine, she said. She wanted my opinion about what we should do.

What the hell did I know about stuff like that?

So I deferred to her judgment. The discussion was way too heavy. Had I not made it clear that two boys were all we needed? And hadn't I clearly heard her agree with me?

I could not imagine a world full of Trojans. We could afford them and Trojans were considered full-proof. Besides, I was thirty-some years old, and as far as I knew there were no little Johnnies running around. I kinda figured I was not a candidate for impregnator.

So I got my first opportunity to do the Trojan thing. I found them to be a nuisance. And, being a nuisance as they were, I became real complacent in their use. In fact, on more than one occasion I was downright derelict in my responsibility to use good Trojan judgment. But who was I to deny my little soldier when he would stand at attention, look up at me with that tear in his eye, and say, "Ah, come on, boss! I don't need no stinking helmet."

Several months later, driving with Teresa on that snow packed street, I heard the word "pregnant" applied in a most personal manner for the first time in my life.

When my mind got back to the road, I noticed the car was headed toward a snowbank. I blinked and made the necessary driving corrections to avoid having to dig the car out of a messy situation.

Pregnant, she'd said. The word scared the hell out of me. Pregnant? I was not supposed to be pregnant. I realized life had just gotten a whole lot more serious. For the first time, I was going to be held personally responsible for my selfish actions.

I had no clue what it meant to be pregnant. But I was about to learn.

By July, I had forgotten about getting that MBA. I had moved my family to Utah and was a full-time working guy. One morning as I was waking up, Teresa came into the bedroom and casually said the contractions had started.

So I casually got up and took a shower, casually got dressed, casually loaded Teresa into our car, and casually started toward the hospital. I casually stopped about halfway to the hospital to get a full tank of gas for the car. When I got back into the car, Teresa casually informed me that the contractions were now four minutes apart.

I panicked. At the hospital, I slid the car into a parking space, jumped out, and ran inside. When I reached the reception desk, I looked around and blurted out, "Where the hell is Teresa?"

Then I remembered. I ran back to the car and retrieved her. She was panting hard and she had some real choice words for me.

About three hours later, my family of two boys had become a family of three boys. By mutual agreement, it was resolved that three was good. My main squeeze and I would never be pregnant again. I was done with Trojans.

Or so I thought. 'Trojan" did come up again, in a slightly different context, but that's another story.

Vietnam
February 1970

When I arrived in Vietnam in February of 1970, I hadn't been in touch with the reality that was America for two years. I had not seen the effects of Tet on the American public. I did not understand the antiwar movement. Worst of all, I did not understand that the Army had changed.

In Vietnam, the Army was moving from an offensive role to a defensive role, from a killer role to a don't-get-killed role. Morale was in the toilet. Drug use was rampant. Blacks didn't like whites. Browns didn't like blacks or whites, and whites didn't like blacks or browns. The situation was normal—all fucked up.

I stepped off the American Airlines jet into an alien world I had no idea existed. I was in the wrong place at the wrong time. There was no room for me, a cocky officer who believed he was invincible. I was not ready or trained for this new world. I had to get trained and get ready, and I had to do it quickly.

I understood that the VC and the NVA were my enemies. I knew that the population was sometimes for us and sometimes for them. I'd been told this, but it hadn't sunk in. I still expected the population to like us once they got to know us—that they would like me once they got to know me.

One day, a week or two after arriving in country, my driver and I were traveling between Cu Chi and Tay Ninh on Highway 22 in a Jeep with a small utility trailer attached. At the south end of a village called Tran Bang, we came upon a sharp curve where we met an ARVN Jeep coming south. As we started to pass each other, a Honda scooter with two young Vietnamese males on it tried to pass the ARVN Jeep and maneuver between it and our Jeep. The scooter hit our Jeep and went down. We ran over the two passengers with the back wheels of the Jeep and the trailer we were pulling. It happened so fast that my driver had no chance to avoid the accident.

By the time we could stop the Jeep and pull it off the road, one of the Vietnamese had already gotten up and he staggered over to a nearby house. The other was sitting up in the middle of the road. Fortunately, there's not much weight over a Jeep's back tires, and the trailer was empty, so the net effect of the accident was that two Vietnamese lost a lot of skin in a very short period of time. I was amazed at how tough they were. If that had been me, I wouldn't have walked away from it.

After we stopped, I got out of the Jeep, told the driver to stay put, and ran back to see how they were. Both of the young Vietnamese men were skinned up a little, but like I said, neither was seriously injured. I had it in mind to offer any assistance I could until I looked at their faces. For the first time, in Vietnam, I experienced what it was like to be hated face to face. They were mad. Not only did they not want my assistance, they didn't want me anywhere close to them.

I sensed that if the power had been theirs, at that moment, they would have punished me severely. I would have paid for this and many things that had happened in their past. I would have been their symbol of vengeance. I saw the same hate in the faces of every one of the Vietnamese who had gathered as witness to the accident.

It became clear that to stand there would be dangerous. I walked back to the Jeep, called Battalion on the radio, and reported the accident.

"Get us out of here," I grunted at my driver.

Take care, my friend. These people not only mean business, but they hate you—personally.

—

Later that day, we were at Alpha Battery to help their maintenance section on some complicated problems with the howitzers. Their howitzers had been in service for a considerable period of time and were in poor shape. During the evening, just before sunset, Alpha Battery Commander Captain Mark Waters asked me to walk with him as he made his nightly perimeter security check. We could use that time to discuss the maintenance problems.

He told me he had been conducting several unannounced shakedown inspections to try to find an AK-47 rifle hidden someplace within the base. He said that he had heard the rifle was to be used to frag the first sergeant and him the next time there was incoming. Fragging was a nasty little term we used to define murder.

I thought nothing more about it until about two weeks later when he called his best firing section out of action. We had decided to send that howitzer to the direct support ordnance battalion for a major overhaul. The howitzer section personnel took several days of in-country R & R.

During the overhaul, the direct support opened the breach of the howitzer and out slid an AK-47. They gave me the AK-47 and I gave it to CPT Waters the next time I saw him.

As I handed it to him, I said, "This is your best howitzer section?"

"Yes, they are."

"What are you going to do about this?"

"I'll let the first sergeant handle it."

Take care, my friend; you may have enemies you never dreamed of.

Oklahoma
Summer 1989

Teresa and I were driving back to Oklahoma from Arkansas, having casual conversations about this and that and generally enjoying the trip.

"How did your conversation on birth control go with Jeremy?" I said. At 16, Jeremy was one of those lucky boys who had females falling all over him, all the time. His age didn't seem to matter. Their age was not an issue, either.

Teresa had come to me a week earlier.

"Jeremy's room is smelling better," she'd said. She had told me on several occasions that you can always tell when a boy starts getting some. Their bedrooms start to smell better.

I could attest to the fact that my sons' bedrooms stank. I am not one to focus on smells, but sometimes I needed a gas mask when I went into their rooms. And then, suddenly, their rooms would start smelling like Old Spice and Aqua Velva.

"You need to talk to Jeremy about the birds and the bees," she'd insisted.

"Why?"

"It's just time that you do it!"

Scared the crap right out of me. "I'm not doing it," I said. "You want him to know about this stuff, you tell him." The truth was, I didn't have a clue what to say to him.

"You wimp!"

"Yep, but I am your wimp."

A week later, driving down the road, Teresa responded to my question. "I went to the store and bought an example of all the different kinds of birth control so I could explain the good and the bad parts to him," she said. "I sat him down and we went through each one in detail. He sat there, very quiet and respectful, and listened to everything I told him. You know what the little stink said when I was done?"

I responded, "No, what?"

"He said, 'Thanks, Mom, but, I prefer Trojans!'"

So, now I have one son and at least four grandchildren thanks to our family's brand loyalty.

Battalion, Cu Chi, Vietnam
February, 1970

Staff meetings were a time for us to hear about the poor quality of our soldiers. We heard about how we needed to double our efforts to get the job done because we weren't getting any support from the ranks. Our battalion could not move, it could not shoot, and it could not communicate, because the troops were no good.

But the troops weren't the problem. The problem was the leadership. Troops can't follow if they have no leaders. And leaders can't lead if they don't know where to go.

I didn't know it yet, but new leaders of the American Army were beginning to emerge. For the most part, they were not officers or senior NCO's. Most of them were just soldiers who could see that the traditional command structure had broken down and left a vacuum of leadership. The opportunity was there for the taking.

So these new leaders started talking.

Fragging was retaliation within the ranks to punish bad leadership. The rank and file could take a lot of crap from bad officers, but at some point they would take action to stop it. Fragging was nothing new. It was found in every war ever fought. In Vietnam, a typical fragging tactic was to put a fragmentation grenade in a can, affix the can to a door, tie a string on the grenade, and tie the other end of the string to the doorway. When the door opened, the string would pull the grenade out of the can, releasing the spoon and activating the grenade. When the grenade exploded, the bad leader would be gone. A fragging here, an assassination there. Rebellion, riots. It was all there in 1970.

Uncle Ho and Sun Tzu would have been proud.

It had become clear to me that there was nothing I could do to resolve the sorry state of maintenance in the battalion. I did not have the training, the trained personnel, the equipment, or the facilities to do the job. There appeared to be no interest within the command structure to correct these problems, so I decided to stop trying and let the chips fall where they may.

I was preparing to take a stand against my first real enemy of the war.

It all began when the battalion commander, Lieutenant Colonel Gray, had ordered me to make him a reproduction of the daily deadline report that I used in the daily staff briefing.

LTC Gray had been in country for two years waiting for a chance at a combat command. He was trying to punch his ticket for general rank. I had been in the battalion for one week when his chance finally came. He was some general's favorite gofer; apparently he had been good at it. In a way, we were both new at the art of artillery battalion combat, and neither of us had been properly prepared for it.

We met with him the day he arrived. He was all starched fatigues, new web gear, polished boots, and hard-ass talk. He laid down the law as to how we, the staff, were to support his command. It was an uncomfortable meeting that set the tone of our relationship from then on.

From that moment, we understood that his perspective of the battalion was that of someone who had a personal need to punch his own ticket. He saw the battalion only from the perspective of higher-level staff. He had no idea what made the battalion tick, and had no apparent desire to learn. His knowledge came from reports that were political in nature and seldom reflected the reality of the unit.

Herein lay my first combat experience with the army that fought and the army that did not. One army's mission was to survive the enemy. The other army's mission was to survive the bureaucrats. We had to balance these two armies. Either one could hurt you.

For the bureaucrats we reported numbers. The reporting structure that had developed within the battalion was a curious sort of counterintelligence. We devised reports that reflected what we thought would make the REMF's (rear area mother fuckers) happy, with just enough reality to allow them to be able to sell it to their REMF bosses. If we were good at devising the report—meaning there was enough truth in the numbers to make them believable from a looking-good point of view—everyone was happy. If we were not, the starched-fatigues bunch got upset. The key was to know your paperwork well enough to trick the REMFs.

Good numbers equaled good reports.

I chose to spend my time fixing the howitzers so they could shoot. It seemed logical to me that if the artillery round landed where it should and when it should, and there was some confidence the next artillery round would do the same, I had done my job. I chose not to worry about reports. I didn't think about the reports. I ignored the other army.

The reproduction of the daily deadline (broken equipment) report LTC Gray ordered was to be small enough to fit inside a pocket of his fatigues, so he could carry it with him as he visited the firing batteries. He wanted it designed so I could change it daily, keeping him up to date as he questioned the battery commanders about their maintenance posture.

All my enlisted men were more importantly engaged, so I designed and produced the reproduction myself. It took most of the day. I suspected LTC Gray wanted the report with him so he could use it against the battery commanders he considered to be underperforming. If they did not clean up their act, he would relieve them from command.

It chafed my butt that I gave him the same information in the daily briefing. Understaffed as the maintenance section was, this additional work fell to me. I resented it. I did not have time for such trivia.

I tried to maintain good working relationships with the battery commanders so that they could be honest with me about their problems. If LTC Gray used my report in a face-to-face confrontation with the battery commanders, it would not take long before I would stop getting honest feedback on their problems.

But LTC Gray was making me kiss his ass. And like I said before, I resented it.

What followed would cripple my career, the careers of several of my peers, and finally, LTC Gray's.

I knew I was in trouble. I didn't know my paperwork well enough to play his game. I knew the battalion was all fucked up in maintenance, but I didn't know how to cover it.

Until that day, the objectives of the staff had been to keep the howitzers moving, shooting, and communicating, no matter what it took. For me, it meant getting the parts to the units as fast as possible, by any means I could. My time was spent doing and reacting, not reporting. I didn't know how to report.

At the staff meeting following LTC Gray's first use of my report, he told me he didn't like it. It didn't fit comfortably into his upper breast pocket, he told me. I had designed it to fit into the lower pocket, the bigger one. But LTC Gray said he didn't want to use the lower pocket. It made the uniform look sloppy, is what he told me.

Then he told me to redo the report.

I left it on his desk after the meeting. I had no further need for it. I didn't have the time or the inclination to do what he'd asked. Leaving it on his desk was a mistake.

The next night he started the staff briefing by announcing that we were to receive two new howitzers the next day. This was news to me—good news, too. At least that's what I thought. When it came time for my portion of the briefing, he told me to send three of my section to Bien Hoa to pick up the howitzers.

Made sense. "Yes, sir," I said. It was our job, after all.

Then he told me to send two mechanics to Bravo Battery. As an afterthought, he added that I was to stay at battalion. So I'd be more available when he needed me.

I did the calculation in my head. Three people to pick up the two howitzers. That would be Specialist 5 Hines, my vehicle-tracked retriever operator, SPC 4 Wilson, my track mechanic, and someone to drive the Jeep—I'd send my parts clerk, SPC 4 Varnes. Then for the two mechanics to go assist at Bravo, I had SPC 4 Pratt, my generator mechanic, and SPC 4 Wilson. But he had to go to Bien Hoa. He was the only one left who was qualified to drive a howitzer. I couldn't do it, since LTC Gray had ordered me to stay behind.

So I spoke up with way too much emotion. "Sir, it ain't going to happen. The Maintenance Section is at less than forty percent staffing. I don't have a maintenance warrant or a senior maintenance sergeant, E7. I don't have enough bodies to send them all over hell. Something's got to change, or we are all going to be fucked in short order."

LTC just sat there glaring at me, and I sat there glaring at him. My outburst took me by surprise. I had never even considered talking to a superior officer like that.

But, damn it, he did not know what was going on in the battalion, and it seemed like he wanted to run the organization like we were in a peacetime army.

He lowered his eyes to his desk and kept them there as I finished my briefing. I knew I had just destroyed my military career.

I felt good, at peace for the first time in a long time.

The voice came back.

That's it, my friend. Your first battle was to learn that you do take up space, even in Vietnam.

—

I stayed at the battalion the next day, as I'd been ordered to do. That afternoon, I received a call from our direct support maintenance company in Bien Hoa. My men had picked up the two new howitzers and hadn't gone more than 15 kilometers when both of the howitzers' motors burned out. The two howitzers were now back at the direct support company awaiting new motors.

I began to laugh. Just a chuckle at first, then a laugh, then a roar. I had to go outside to settle down. The rest of the staff looked at me as if I had gone *dinky dau*. Crazy.

They asked what was funny, but I wouldn't answer. I was not even sure I knew. Maybe it was the battalion, maybe it was how the battalion couldn't do anything right. Who else could draw two new howitzers and destroy them before firing a single round? Who else but me, I thought. What a joke.

The solution to the lack of manpower was to call the battery commander who was to receive the two new howitzers and ask him to use his men to replace the howitzers in Bien Hoa. We reasoned that if we sent the people who would be using the equipment, and who were already trained in the operation of the equipment, nothing should go wrong.

The operator's manual states, in language designed to be understood by fifth graders, that you are to check the fluid levels in the howitzers engine before you start it. The oil and the water and the hydraulic fluid. If these fluid levels are low, you are to fill the reservoir. Then the manual states that when proper fluid levels are established, you may start the motor and drive off.

I understood it, so I assumed that these three would understand it, too.

I learned from the telephone call that our soldiers signed for the new howitzers, jumped into them, started them up, and left for home. There were no pre-operation checks.

There was no coolant in the radiators and no oil in the motor. The expected course of events took place. The motors burnt up. More important, these three soldiers got an unexpected in-country R & R. These guys were going to get a few days off while I got my ass chewed.

I had to give it to them, it was brilliant. An act of true leadership.

I laughed until my sides hurt. It was typical of this unit—true leadership was manifest in anyone who had the will to exert it. The officers had long since given up that position. Their only objective was to survive their year, collect their rubber-stamp medals, and go home. No one cared how the war was going; few ever took the time to check.

After my outburst at the last staff meeting, I figured I had made the list of officers that LTC Gray would relieve from duty. If I was going to go, I might as well go out with a bang. The bang was going to come that evening at the daily staff meeting. I had the ammunition I needed to defend myself. Now I could show the dumb bastard what I was talking about.

The briefing started with a certain air of anticipation. Everyone in the room knew it was my turn in the barrel, and they were hungry for a good bloodletting.

When it came my turn, I uncovered the deadline report which showed the two brand-new and already broken-down howitzers.

"What are you trying to pull here, Captain?" LTC Gray said. "These numbers don't add up." There was venom in his voice.

"We didn't get the two howitzers. They are deadlined, sir."

"What do you mean, deadlined?"

"Do you want the long or the short of it, sir?"

"Just give it to me straight, Captain."

"OK, sir. Those hot dogs we sent to Bien Hoa burnt up the motors in both the new howitzers. Both units are now back at Direct Support waiting for new motors."

"What do you mean, Captain?"

"It appears, sir, that these hot dogs either could not read, did not read, did not want to read, forgot, or intentionally overlooked the operator's manual. The net result was that no pre-operational checks were made. The howitzers were started and moved without any oil or coolant in the motors. This caused the motors to burn up."

"Sir, maybe if someone from Battalion Maintenance had been with them, this wouldn't have happened."

Our eyes met and I refused to look away. His face said, "Why, this insubordinate son-of-a-bitch," but his mouth remained silent until he looked down at the pocket-sized deadline report that I had left on his desk. He asked me if it was up to date.

"No, Sir."

"Why not, Captain?"

"Because you don't like the size. It makes your uniform look sloppy. I have not had time to redesign it. Why should I waste my time keeping that up to date if you won't use it?"

"Then get the damn thing off my desk!"

He threw the report at me. It landed on the floor about halfway between us and I just let it lay there. We stared at each other for a few seconds. I let him win that staredown. When I looked away at the other staff members, I may have been smiling.

Take care, my friend. Do not forget what you are about here.

Later the same evening, LTC Gray called me into his office. I knew it was time for the axe to fall, but I did not expect what happened. After I took my seat, he threw the quarterly maintenance report at me and asked if I had intentionally falsified it.

The quarterly maintenance report was the bureaucratic numbers game that went up the chain of command. This report justified everything that was American in our battalion in Vietnam. It was supposed to be a recap of how every piece of equipment had operated for the prior three months. I was taken aback. I had expected anything, including a duel, but I had not expected him to question my integrity.

Luckily for me, I hadn't been at Battalion when the report was done. I'd been at Fire Support Base Hull. The Battalion XO had completed the report, signed my name, and sent it in.

It was one of those political reports that took a lot of know-how to falsify. Unfortunately, the XO had done his best to tell the truth. This made everyone mad because it screwed up the paperwork all the way up the line. I'd had a chance to skim the report about a week after it had gone in. I could tell it was inaccurate, but since I didn't care one way or the other, I didn't think to try to change it.

"Captain, are you responsible for this report?"

"Yes, sir."

"Is this your signature?"

"No, sir."

"It's your name. If you didn't sign it, who did?"

I'd just seen the XO who'd made the report. Major Smith, in his office next to LTC Gray's. I knew MAJ Smith could hear everything being said and figured he was nervously waiting for my response.

"I don't know who signed it, Sir," I said. "I was not here when it was made out."

"It appears, Captain that you intentionally tried to falsify an official government document."

He was trying to court martial me. I knew I had to be very careful with every word that came out of my mouth.

"Is it wrong, sir? I have never read it, sir."

"What do you mean, you have never read it?"

"It was done when I was not here. I have been so busy that I forgot all about it when I got back to Battalion. If it is wrong, it is my fault, but you know more about it than I do."

He sat there looking at me for a few moments and finally said:

"So, you never signed it?"

"No, sir."

"And you don't know who did? Did MAJ Smith sign it for you?"

He knew Smith had signed it. He wasn't that stupid.

"I doubt it, sir." I hoped the major heard me.

"Who do you think signed it, Captain?"

"I wouldn't tell you even if I knew, sir."

He let several moments slip by. Finally he said, "I have been levied for a captain because of the losses at Fire Support Ellsworth. I've decided that I'm going to send you. You have clearly demonstrated that you cannot work for me. You cannot be effective in my command, and you are a threat to my authority as long as you are around."

I was stunned. I'd won. He was transferring me out of his worthless outfit. Fuck up and move up. It really worked.

All I could think to do was smile. I must have looked like some kind of jerk. I was so happy I could have kissed his ass.

Well done, my friend. You have met and destroyed your first enemy.

The Group Commander, Colonel Lindell, made me wait two days before he had the S1 notify me to report to him. No one talked to me during that time. I was stashed in a bunker by myself. I didn't mind being alone for a change. I used the time to read a book. It felt good not having to deal with the pressures of the last month.

I felt like I might live after all.

The meeting was going to be very unpleasant; COL Lindell had a reputation of being hard. The only times I had seen him, in his starched fatigues, polished boots and clean web gear, was when he was getting in or out of his command-and-control ship. He always looked well fed and clean, and I resented it. He always looked so self-important, and I resented it. The only times he had ever talked to me were to give me orders, and I resented it. He always had gofers around him and they always looked just like him, and I resented it.

This meeting was not going to be easy on me. I had rebelled against the leadership of a superior officer in front of other officers. This was an unforgivable sin. There was not enough evidence to bring me up on charges, but that was not necessary. The military system would not tolerate insubordination in any form. COL Lindell had the power and the means to punish me for my indiscretions. I was about to learn my punishment. He would not have been much of a commander if he looked the other way after what I had done to one of his officers.

I was prepared to receive his wrath.

In a way, I guess, I looked forward to the punishment I expected to receive. I needed it to clear my guilt. I had committed the cardinal sin within the brotherhood; I had openly questioned a superior officer and shown signs that I was no longer willing to follow his leadership. To do it in private is bad, but to do it in front of my peer group was unpardonable. I had chosen the latter course. Now it was time for me to pay.

COL Lindell made me stand at attention for a while and did not return my salute for what seemed like an hour.

"You look like a whipped dog," he finally said.

I felt like a whipped dog. As much as I'd savored my triumph at the time, I still wanted to be a good officer.

"LTC Gray reported to me about what happened. Do you have anything to say for yourself?"

"No, Sir."

I wanted to say that I didn't volunteer for Vietnam to become a fuck up but I thought better of it.

He proceeded to lecture me on the brotherhood, my obligations to it, and my responsibilities relating thereto. I didn't pay close attention and my mind wandered. Then I heard him say that as long as I was being paid by the Army, I would do as I was told.

When he said that, my mind reacted.

Did he just tell me that I am a mercenary?

I didn't tell him I'd come to Vietnam because it was my duty, not because the Army signs my paychecks. I didn't tell him that if he couldn't understand that, then we would have a problem. I didn't call him an asshole. But I did think these things.

I don't remember another thing he said.

His correlation between my sense of duty and my paycheck was the seed of a larger problem. I had taken an oath as an officer in the United States Army. My pledge was to the Constitution that existed as a manifestation of the established military system. My philosophy had been that my pledge and my belief in the system would be sufficient to motivate my activities in service to my country. This philosophy failed in that room in 1970 because it was false logic.

I was standing before a man who equated my sacrifice, my service, in terms of dollars. He considered me a mercenary.

How revolting.

Take care, my friend, take care.

Ogden, Utah
1978

Home was Teresa, Matthew, Jeremy and Charles, dogs and cats and funny-looking little fish, birds and turtles and snakes and bugs. Boy Scouts and show time in school and other assorted good things were to follow.

One day I came home from work and Matthew met me at the door. Some idiot, probably some kid's mom, had had a brilliant idea for a fundraiser.

It was decided that a good father/son activity in the Boy Scouts would be to bake a cake together. Matt, my Boy Scout, had never baked a cake, and he didn't know where to start. Neither did I. Even more important, the male in me insisted that men don't bake cakes.

So we enlisted Teresa to consult with us. The project immediately transformed from the act of baking a cake to the production of a piece of art.

We should have kept our mouths shut and just did the best we could. But no, we had to enlist Teresa, who'd never be seen in public with just any sort of cake.

Down to the store we went. I bought a white cake mix and two colors of canned icing. Matt and I baked the cake, which somehow turned into a good depiction of a Snoopy dog. I have to admit it didn't look all that bad once we had finished with the icing.

So, off we went to the fundraiser, Matt in the lead with his state-of-the-art Snoopy cake. Proud Momma, reluctant Dad, and two little brothers in tow.

There were already thirty or forty cakes on the table in front and more than a hundred people in the room. I had to admit, compared to the other cakes, Snoopy Dog was looking good. About a third of the folks present were reluctant dads, just like me.

Matt said, "Dad, can I bid on our cake?"

"Yeah, go ahead."

Matt ran off with his friends to the other side of the room. And the bidding started. About half of the cakes had been sold when Snoopy came up. The average winning bid was about ten dollars. I figured Snoopy was worth ten bucks.

The Boy Scout troop leader began the bidding. "So how much are we bid for Snoopy?"

"Five to open."

Matt: "Ten!"

Some jerk (a friend of mine): "Fifteen."

Matt: "Twenty!"

Some other jerk (my best friend): "Twenty-five."

Matt: "Thirty!"

Jerk number two: "Thirty-five."

I yelled at Matt to stop and started to run across the room.

Everybody laughed and did what they could to slow me down. By the time I could get Matt shut down, that damn dog had cost me fifty bucks.

And I don't even like white cake.

Tay Ninh Province
March 1970

Tay Ninh Province was spooky.

This would be a hell of a place to die, I told myself as I looked up at an eerie glow burning a circle into the sky at the top of Nui Ba Din, The Black Virgin Mountain, just north of Fire Support Base (FSB) Hull. I wondered if the halo was for me.

FSB Hull contained Alpha Battery. It was one of two fire support bases in the neighborhood of Nui Ba Dinh. We had firing batteries in both FSBs. I was spending most of my time at one or the other

FSBs working on maintenance problems. We could tell something was up, so maintenance of the howitzers had become a priority.

We were preparing for something big. Whatever it was, it was being planned by those far above my pay grade and I hadn't been made privy to the details. Of the two batteries, FSB Hull was the worst off with two howitzers out of commission and immovable.

Consequently, I spent many nights at FSB Hull, staring up at that white halo.

Nui Ba Din was not so much a mountain as just a large pile of rocks that stuck out of the plains in War Zone C just northeast of Tay Ninh City. It dominated the terrain for hundreds of square kilometers in all directions. We had a communications base secured with extensive defensive constructions at the top. The defensive lights at the perimeter of the communications base created the halo.

The rest of the mountain belonged to the VC and NVA, and their territory was undisputed. It was a strange situation. There were stories that we shared the same waterholes with the enemy outside the perimeter. We had tried to clear the mountain several times and sustained heavy casualties without making any headway. Air observers had seen enemy tanks on the mountain but could not engage them effectively because of the rocks. You couldn't fly close to the mountain without risking being shot down. Over the years, a sort of truce had developed. We left the middle of the mountain alone, and they left our retransmission station alone at the top.

Alpha Battery and Fire Support Base Hull, (FSB Hull) were located south of Tay Ninh City in a farming area situated close to two small hamlets, one just north and west of the FSB and the other just south. Highway 22 ran along the west side of the FSB. Just west of the highway was the Vam Co Dong River.

The only free fire zone existed to the north and northeast of FSB Hull. A free fire zone was an area under the control of the enemy and we were free to fire any weapon at any time.

The perimeter of FSB Hull had been cleared out to about a hundred meters on all sides. The north to southeast sides faced rice paddies and there were clear zones of fire out to about two thousand meters.

FSB Hull was circular in configuration with a dirt berm surrounding it. The defensive bunkers were built into the dirt berm. Outside the berm were three barriers of concertina wire interlaced with Claymore mines, trip flares, and 50-gallon drums of phougas.

Phougas was a nasty little thing when it worked. It was napalm in a 50-gallon drum fixed with a Claymore mine. When the Claymore mine was set off, the ignited phougas mixed with the Claymore shot and it toasted or ripped apart everything in the kill zone. The result kind of reminded me of overdone hamburger.

All key installations in the FSB were in bunkers. The fire direction center (FDC) was the nerve center of the FSB and it was occupied at all times, good and bad.

The living quarters were in bunkers. These bunkers were constructed of heavy timber and three layers of sandbags. They'd been designed for protection from rocket and mortar fire, but they were not designed to be occupied during a ground attack.

While FSB Washington, the other fire support base by Nui Ba Dinh, was secured by a platoon from the 25Th Infantry Division, FSB Hull was secured by a platoon of the Army of the Republic of Vietnam Infantry. This was part of President Nixon's Vietnamization program. The platoon-sized ARVN unit was commanded by a lieutenant who was supposed to speak English. I met him on the day Captain Waters and I inspected the perimeter. He spoke about as much English as I spoke Vietnamese.

We kept trying to tell him that the det cord he used to set up the Claymores and the 50-gallon drums of phougas were no good. It had been out in the weather too long and needed to be replaced. The lieutenant would just look at us, smile, shake his head as if he understood, and then he'd go about his business. Come to think of it, even if he could understand us, he probably couldn't have done anything about it since he probably didn't have any det cord or any way of getting his hands on some.

—

The American Army of 1970 was more comfortable in the daytime. Charlie owned the night. We spent the days moving, shooting and communicating, when it was Charlie's turn to hunker down.

During the night we hunkered down and prepared for Charlie to move, shoot and communicate. We had several rituals for the transition from day to night. One particularly spooky ritual at FSB Hull was the "Mad Minute."

The Mad Minute was an all-out weapons test by the ARVN security force and the American firing battery personnel. Every few days just about sundown, everyone would be directed to test their weapons, everything from a .50 caliber machine gun to a .45 caliber pistol. To make sure their assigned weapon of defense was working properly, everyone was directed to fire as many rounds as possible in one minute.

The Mad Minute was meant to allow the weapons operators a chance to check their weapons, but FSB Hull's two no-fire zones meant that many of their weapons never got tested. What was worse, the enemy knew where the no-fire zones were and could exploit those areas to their benefit.

As we would soon discover.

Late one evening, Captain Mark Waters and I were walking around the FSB when we heard something outside the wire. It sounded like metal hitting metal, as if someone was pounding a stake into the ground. The pounding would stop for a while and then start up again.

We crawled onto the top of the Fire Direction Center bunker and took a starlight scope to see what we could see. The starlight scope used ambient light in the night to create a green, magnified image. Walters looked through the scope and said: "There he is."

Mark handed me the scope and pointed in the direction he wanted me to look. I raised the scope to my eye and looked through it at the green-lit image. There he was, a young man, probably a sapper, dressed only in black shorts and sandals, holding a hammer in one hand and a metal stake in the other, standing on top of a dirt mound a hundred meters outside the wire and looking directly at us.

He was between us and the hamlet to the south, so he knew we would not shoot at him. He just stood there taunting us, daring us to shoot at him. We watched him for a while and decided he was conducting a survey, probably getting ready for a mortar attack.

Mark told me they'd seen men in the trees around there for several nights in a row, and that his repeated requests to fire on them had been denied.

That night, we shot a flare towards the sapper. The flare popped about 100 meters over the spot and hung under its parachute, swinging back and forth. It cast a lazy moving light over the area with shadows coming and going. Charlie stopped what he was doing and moved behind a pile of dirt where we couldn't see him. We watched until the flare burned out and hit the ground then we crawled down off the FDC and continued walking around the FSB.

Several weeks later, during our invasion of Cambodia, elements of the 11th Armored Cavalry Regiment would discover the enemy's field survey of FSB Hull, the one they used for the attack on the fire support base. It varied from our own survey for Battery Center by one meter to the north and one meter to the east. They were right on. Our friend did a good job.

Oklahoma
1980

Five-year-old Charles had decided to become a bone collector. He stored them under his bed. One day we were out in the country and Charles filled up the trunk of the car with old bones. When we got home, we had most of the skeleton of a "cowse." I had never seen a skeleton of a "cowse" before. Until then, I didn't even know there was such a thing as cowses in the New World.

Cowses are a cross between a horse and a cow. I had never seen a horse with horns or a cow with horse shoes. But cowses had both. And Charles could prove it to you, if you weren't scared to look under his bed.

So Wilbur, Charles' cowse, went to show and tell at school. And then Wilbur lay under Charles' bed for years. One day we moved, and Wilbur got lost. I suspect that Teresa found him a comfortable place in the garbage can.

Years later, Charles reached middle school and Matt left for college. While in the dorms, Matt built a significant library of *Playboy* magazines and moved his library home during winter break. When he moved back into the dorms, he left his collection hidden in his old room at the house, safe and sound.

When Matt returned for his *Playboys*, he discovered that his whole collection had disappeared.

Matt was very possessive of his things. We all knew we were in dangerous waters when we messed with his stuff. Charles, on the other hand, knew that messing with Matt's stuff was the easiest way to pull his chain. Charles was an expert at pulling Matt's chain.

Naturally, Matt assumed his mom had thrown his library out. He lost it. Teresa related it all to me later. It went something like this:

"Mom, what did you do with my *Playboys*?"

"Nothing."

"They're missing. Did Dad throw them out?"

"I don't think so. He probably didn't even know you had them. He probably didn't care one way or the other. Have you talked to Jeremy or Charles?"

"Not yet."

"Charles is upstairs in his room. Go ask him."

Matt charged into Charles' room with Mom right behind him. She was only there to protect Charles from a potential thrashing. Charles was lying on his bed reading a comic book.

"Charles, do you know what happened to my *Playboys*?"

"Yep."

"Where are they?"

"I sold them."

"You sold them? To who?"

"I took them to school and sold them to my friends for twenty bucks apiece."

You got to love those boys!

Fire Support Base Hull
March 1970

I was dreaming again of the dead NVA soldier on the pile of bodies at Cu Chi. He began to move. Slowly, first his arm, then one leg, then the other. He sat up on the pile of bodies, turned, and looked straight at me. His eyelids were closed, but I could see his eyes anyway. They were on fire. He smiled at me.

It's a bitch, ain't it, friend? Well, it's 4:45. Time to get up, time's a-wasting. Today's the day, sport. If you pass, you're in. If you fail, there won't be another chance. Today you join the society or you don't. It all depends on how you do today. So get your ass up.

For some reason I felt more tired than usual. I wanted to sleep more.

Look, asshole, if you don't start moving, and I mean now, I'm going to jerk you up by the stacking swivel.

Then he looked down and addressed the other bodies in the pile.

Who does this asshole think he is, anyway? When I say jump, he's going to learn to ask "how high?"

I was sleeping soundly in my cot when FSB Hull was struck by a three-pronged attack.

My cot was in the officer's bunker, where there was no dress right dress—that's military for keeping things organized and in order. We'd arranged things at random so it would be harder for a sapper to find a sleeping soldier's throat to cut. It was 1970, remember; by then we'd had considerable experience with the futility of lining everything up in neat rows.

My cot was in the center of the bunker. It was raining hard. The three layers of sandbags that covered the bunker would absorb a lot of water, but once they were saturated, the inside of the bunker wasn't much different from being outside in the storm.

A stream of water broke through the sandbags and rained down on the small of my back. I thought about moving my cot, but I was so tired. The stress of trying to do a job I was no good at coupled with the stress caused by the risk of living in wartime Vietnam, along with long work days and nights and marginal opportunities to eat wore me out real fast.

I did not want to make the effort to move my cot. Besides, I reasoned that if I moved there would probably be another stream of water anyway. So I decided to ignore the water.

Someone came in the bunker about midnight, saw the water hitting me in the back, and moved my cot with me on it. I appreciated it, but not enough to wake up and say "Thanks."

Everyone settled down and someone turned out the one light bulb.

Sometime later, I heard yelling. I thought I was dreaming for a while, but the yelling didn't stop.

A voice inside the bunker yelled. "Goddamn, son of a bitch, who the hell is doing all that yelling outside the bunker? It is too fucking early."

"Now, what?" another voice yelled. "What the fuck are the ARVN shooting at? When are those bastards going to get their shit together?"

"We're under attack!" a third voice yelled.

All I wanted to do is sleep. Flashlights with red filters came on, I could see Mark Waters sitting on his cot putting his feet into his boots. He stood up, grabbed his web gear, M-16 and helmet, barked orders at his executive officer, and ran out the door.

I thought this was their problem, not mine, and I tried to go back to sleep.

The XO (Executive Officer) got up and left just as several explosions from incoming mortars hit nearby.

Until the explosions, all I wanted to do was sleep. But then my mind started to work. *Holy shit! Wake up, you dumb fuck! Get your ass in gear, this is no time for wet dreams!*

So I woke up, sat up, put my feet into my boots, found my helmet and web gear, and stumbled toward the bunker exit.

All I could find in the dark was my .45 pistol and two clips of ammunition. I ran out the door with unbuttoned pants, untied boots and my .45 that I had never fired. I looked around the FSB and noticed that the most activity was to my right and on the berm. So I ran and jumped up on top of the berm.

"No shit," I called to no one in particular. "There they are—sappers, and they are in the wire."

I let fly several rounds from my .45, then just stopped and watched what was happening. At that moment I was aware of the mortar rounds and the sappers.

We found out later that the VC were using mortars, sappers, and a platoon of infantry. It was a well-coordinated attack with mortar rounds raking across the base from north to south. The sappers came through the wire from the northwest, shielded by our no-fire zone.

The platoon of infantry came in from the east. They were protected by the shitter, a three-holer that the battalion doctor had insisted be placed outside the berm for health reasons. The shitter was between the berm and the wire. It effectively screened a width of wire wide enough to drive a deuce-and-a-half through from our observation.

No one on our side had thought to check the field of fire and observation after the three-holer went up.

As I stood on the berm, I watched the carpet of tracers going out and the flicker light of the flares causing everything to dance around. I could see a body out in the wire nearly disintegrating as the 60s ripped into it.

Shit, man, I thought. *This is a gas.* What a way to fight a war. I looked up at the halo in the sky and yelled, "Not tonight, motherfucker, not tonight!"

It was a well-designed attack, with the sappers given just enough time to cut through the three wire barriers before the mortars started. When the mortars started, the sappers would then charge the center of the fire support base with satchel charges and eliminate the key installations, such as the fire direction center and as many sleeping bunkers as possible. They were to attack and destroy as many defensive positions on the east berm as possible. This would allow the platoon to attack with minor resistance and destroy the base.

The sappers had already reconnoitered the wire and knew where each flare was located. They would crawl up to the flare, grab it off the wire, slip a rubber band around it to keep the spoon from flying off, and hand it back to a buddy behind them.

A single slip unraveled their beautifully orchestrated attack.

While cutting through the third and final barrier of concertina wire, a sapper passed the last flare back to his buddy. But his buddy missed the handoff and dropped the flare. It fell to the ground and the spoon flew off, igniting the flare. When it went off it exposed their position.

All our defensive positions opened up with everything they had. I stood on the berm and watched VC getting shredded as they attempted to retreat back out of the wire. The .50s shot a few times until they jammed, the .60s worked well, and the M-16s shot all over hell. There were M-1 carbines, .45 pistols, mortars and M109A1 howitzers all going off all at once. The carpet of tracers pounding into the sappers was so thick, you could damn near walk on it.

It was beautiful.

The sappers who were still alive got up and ran off. We killed two, got one blood trail, and missed the rest. The distance from the rifles to the sappers was about fifty meters when the fun started. I was amazed at how bad our aim was.

When the VC platoon leader realized that the sappers had failed, he called off the attack and retreated. We didn't even know the platoon had been there until the next morning, when we discovered the enormous gap they'd cut into the wire behind the shitter.

Bad aim and a dropped flare weren't the only misfires that night. We found that not a single Claymore mine detonated. The reason? Defective det cord.

The .50s did not work because the ARVN troops using them did not know how to set the headspace, which is the distance from the barrel to the firing mechanism. Nor did they set the timing of the weapon. If the headspace is wrong, the shell won't seat properly before firing. If the timing is off, the firing pin won't strike the primer. Either situation will render the .50 non-operational after about three or four rounds.

The whole affair was one of pure luck—incredibly good luck on our side, and unbelievably bad luck for the VC. Not only had the attack failed because of one dropped flare, but the VC platoon retiring from the battle stumbled into an ARVN ambush and was destroyed. Had it not been for one dropped flare, they would have jerked our strings.

The tactics of the losers could not be faulted, and the tactics of the winners were nonexistent, but the winners still won and the losers still lost. The battle for FSB Hull was over.

If the attack had been successful, it would have been a good propaganda tool against Vietnamization. Vietnamization, whatever that was, had worked.

Now that I think about it, what a crazy war it was, that war of ours.

Oklahoma City, 1985

The dream was back.

I had been running from the VC for hours. They were trying to recapture me and put me back in the cage. When I could no longer sense them behind me, I stopped. I needed to rest. I climbed under a bush and laid down. The terror of losing my freedom again was real. I feared I was going to lose my self-control any second. It seemed like the best I could do was move in slow motion. No matter how hard I tried, my body just slowed down. It was hard to move my arms and legs. It was hard to breathe.

I decided to take a chance and look down the hill to see how close they were.

Damn it, I thought. A young VC was walking straight toward me. *Settle down, asshole. Don't breathe, he might hear you.*

When the young VC got close enough, I jumped up and ran at him. My body and legs moving slow, I feared I could not get to him in time. Then, somehow, I got there. I hit him before he could raise his rifle and we went sprawling back down the hill. As we rolled, I slammed the knife as hard as I could into his chest. I felt the blade glance off a rib and then slice through the muscle as it plunged deeper into his body. I gave the knife a couple violent jerks so the blade would rip into organs, shred blood vessels, and open up a hole in his lungs so the blood would drown out his pleas for mercy.

For a second I just lay there on top of him with one hand over his mouth so he couldn't scream. I looked into his eyes, and he looked into mine as if to ask why I had hurt him so. I didn't want to watch him, but I couldn't look away. I felt his body relax as I watched his eyes slowly lose interest in me. It looked like he was trying to look at something behind me, a long way off, up in the sky.

I got up and looked at my hands, dripping blood. I tried to wipe it off on my fatigues, but it would not come off. The blood kept dripping. Drip, drip, drip.

Then I heard the chopper coming to get me out of there. The other VC were far enough away that I knew the chopper would reach me first. It hovered over the trees and a crew member dropped a rope ladder. I grabbed it and hung on as it lifted me off the ground.

I looked down and saw hands and arms close around my legs. The knife was still in his chest. That young VC's eyes were dead, but he was not. I screamed and kicked as hard as I could.

Oh, God, I thought. *I am not going to escape Vietnam.*

Teresa's shouts jolted me from my dream.

"Ow!" she said. "Damn, John that hurt!"

Coming to my senses in our dark bedroom, I said, "Where are you?"

"Over here on the floor. You kicked me again."

I was so embarrassed I wanted to cry.

Teresa walked toward the bathroom. She mumbled something about being tired and having to get up early for work, then slammed the door behind her.

After ten years, she'd come to recognize the conditions that resulted in the nightmares. Usually they were predictable—something I read or saw on TV would result in a bad night. She'd learned when and how to break into the nightmare. Most of the time all she had to do was touch me and say: "John, you're dreaming."

She knew that once the dream had been broken I would settle down and sleep quietly. I just had to remember I was safe.

God had sent me a guardian angel with the chore to change my nightmares into opportunities. It was a chore she willingly took on. Isn't that what guardian angels do? Still, at times that chore was a more than she'd bargained for.

"Damn his hide," I heard her grumble through the bathroom door. "Enough is enough. I really am getting tired of this."

A couple of days later she told me—she didn't ask me—to start writing about Vietnam.

III Corps, South Vietnam
March 1970

During training at Ft. Sill, our instructors emphasized how difficult it was to know who your enemy might be. There were stories about little children throwing grenades into your vehicle just after you had given them candy. There were stories about the *mama sans* selling you iced drinks laced with glass slivers. There were stories of old men sitting on the side of a road and watching you drive or walk by. When you got into the killing zone the old man would touch two wires together, detonating a mine. There were stories about young women intentionally infected with VD and their mission was to pass it on to you. Some strains of VD were incurable; if you got one of those you would be sent to some quarantined place outside the United States and never be allowed to go home.

When you asked the instructor, "What the hell were you supposed to do, when the little ones are as deadly as the big ones?" The instructor would most likely respond, "It sounds worse than it actually is." And he would smile and move on with the training plan.

The training was good for me from one point of view. I made the decision that I was going to be one celibate motherfucker while I was in Vietnam.

For me, the training left me with a few big questions: Who do you kill and who don't you kill? When do you kill and when don't you kill? Do you kill first and ask questions later, or do you wait until you are sure it is kill-or-be-killed time? Is it time to kill when you feel the bullet ripping through your body?

One question never crossed my mind during training, and it never came up when I arrived in Vietnam:

Why kill?

During those first few weeks in Vietnam, I spent almost every day traveling the roads in III Corps Area of Operations, from Cu Chi to Tay Ninh and places all around Nui Ba Dinh. Sometimes the roads were secure, sometimes not. I was always in a Jeep with a driver and sometimes with another passenger.

I took it upon myself to be the security for our little unit. I would make sure there were no grenades, no mines in the road, and no ambushes. Only one time did I discover a Claymore mine set up next to the road we were traveling. The kill zone for the Claymore was the road. I noticed the Claymore as we drove through the kill zone.

Nothing happened.

I had my driver move the Jeep down the road out of the kill zone and stop so we could investigate the Claymore. We discovered that it was wired to a command-type ignition system, so we cut the wire and left the mine in place. Then we un-assed the area. When we drove by the next day, it was gone.

In the villages, I kept an eye on as many of the people as I could. In the country, I watched out for traps and mines. At least I did my best. But there was no way to take it all in. I was merely driving myself *dinky dau*.

My stomach began to hurt all the time. I thought it was because I seldom got more than one meal a day, usually C rations. But it was not a lack of food that was hurting me, it was my nerves. I was trying to be the do-all, be-all guy who would save us from the evil all around. It wasn't working. I couldn't do it all.

So I gave up trying to do the security thing. I decided I'd take whatever came my way. It was hard. It went against my centralist view of the world, where I had complete control over me.

When I made this decision, I relaxed a little, and as we drove down an insecure dirt road, with the red dust coating my face and me kind of enjoying the scenery for the first time, I heard a voice in the back of my mind.

My friend, it said, *you are going to have to realize that a certain amount of what is going to happen to you cannot be controlled. Most of your security will depend on plain dumb luck. You might as well sit back and enjoy the ride.*

So I did.

I had been in Vietnam about one month. It was enough time for me to realize that I was not up to the tasks required of me. Worse, I had no idea about how to get where I needed to be so I could be effective. The problems of the battalion were so large, it seemed there were no solutions. As battalion motor officer, I had been doing nothing but going from one emergency to another; the situation seemed to be getting worse, not better. Our battalion equipment was either worn out, poorly maintained, or naturally defective for the mission we were assigned. Just trying to keep the eighteen howitzers firing was a next-to-impossible task. There was no time to rest. There was no time to plan. There were many days when there was no time to eat. In that short period, I lost so much weight that my fatigues no longer fit.

One night, I found myself at our forward tactical operations center located in Tay Ninh West. After I had settled my men in for the night, there was no place left for me to sleep. I finally found an old abandoned bunker and a cot with no blankets. There was no electricity in the bunker, and it was darker than the hubs of hell.

I considered sleeping outside but dismissed that idea because of the high probability of rain.

The night before, several sappers had crawled through the wire and into the base. They had gotten into some sleeping bunkers and cut the throats of several soldiers. Two of the sappers had been captured during the day, but three others were still at large inside the base. I didn't like the idea of being alone in that bunker with the sappers around. And I didn't like being in a bunker that hadn't been inhabited for some period of time, either. No telling what creepy, crawly things might have been living in there.

I was tired, I was hungry, and I was in a place with who knows who or what. Being alone in that bunker with no lights made me very nervous. But I was so tired. Sleep was inevitable.

I took off my shirt and rolled it up for a pillow. I unbuttoned my pants and loosened my boot laces, but I left them on. I laid down on the cot, my M-16 on the floor beside me and my .45 under my shirt. I listened for any movement for a few moments, then I dozed off.

As I lay there in that strange zone of sleep where many of my senses were not asleep, where I remained alert to any strange noise or a feeling of movement, I felt a cockroach crawl onto my bare stomach. I brushed it off with a sweep of my hand.

Sometime later it was back, crawling toward my face. I brushed it off again.

"Look, asshole," I said. "Leave me alone. I'm leaving in the morning. You can have this whole dump back for yourself. Don't screw with me."

Later, the cockroach came back.

"Goddamn it!" I said. "I asked you to leave me alone!" Then I grabbed it and squeezed until I felt it pop. "You just wouldn't leave me alone, so fuck you!"

Satisfied that I had won that battle, I drifted back into a fitful sleep.

The night moved on and I began dreaming again.

I was alone, in a room, lying on a big bed. The walls and the ceiling were painted white. Nothing hung on the walls. There were no windows or doors. As I stared at the empty walls, they began to move away. The room was expanding, all four walls and the ceiling moving farther and farther away from me. Finally I found myself floating in a white void.

I'd had this dream every so often since I was a baby, mostly when I was sick. But this time it was different. My chest felt wrong, like there was a weight on it. *Goddammnit,* I thought. I couldn't breathe. I thought I was dying.

Desperately trying to claw my way out of the nothingness, I brought my arms up to my chest and found a massive rat lying there. I grabbed it and threw it off as hard as I could. I heard him hit the side of the bunker. I lay there, my body ridged from fear and my open eyes staring straight up into the blackness of the bunker.

A memory from my childhood emerged. When I was five years old, my brother and sister had teased me into jumping into the big irrigation ditch at the ranch. I was afraid of the ditch, but I didn't want them to know or they would laugh. So I ran and jumped in. I didn't know how to swim, and the current was too strong for me to get my feet underneath me. I was suspended, nothing but water all around me, and I knew I was going to die.

As I lay on the cot and stared into the black, I heard:

Take care, my friend. You may be your own worst enemy...

The next morning, I was in the back of a deuce-and-a-half loading it before we could continue on to Alpha Battery. Suddenly a mortar round exploded across the airstrip. I flinched and looked at the cloud of smoke and dust left by the explosion. Time slowed down for me as I looked back at my men. In a kind of slow motion they took off running for the bunker, but I just stood in the back of the deuce-and-a-half. I turned back to the cloud and watched. I felt I was frozen in time. In succession, the second mortar round landed on the other side of the airstrip, but closer. The third round landed on this side of the airstrip, still too far off to hurt me, but it knocked down several soldiers when it went off, covering them in thick dust and dirt. I couldn't see anything for a while. I just stood there mesmerized.

The next round would be closer, I thought. The next one would be personal. I held my breath, closed my eyes and waited. I was ready to go.

But there were no more incoming mortar rounds. A senseless disappointment came over me as I stood and waited. Nothing happened.

But then something did. In what seemed like a lifetime but could only have been a few seconds, a part of me died and another part was born, both destroying and creating, opening new horizons and closing old realities. The old John, the one my family and friends had known, was dead. The new John, the alive John, was something at once new and unpredictable.

I watched as a few of the soldiers were getting up, dusting themselves off and limping off toward some buildings by the airstrip. Others lay still.

Look at that, my friend. Some of those boys ain't getting up. They must be hurt bad or dead. What the fuck, they're not your men, it really isn't any of your business. Besides, you need to get this truck loaded. Let's move it!

And that is exactly what I did. I felt calm. I entertained the idea of going crazy. It seemed like it might be pleasant to just let go of this fucked up reality.

But how could I be sure I'd ever make it back from insanity? How do I un-insane myself? I knew there was no guarantee. So instead, I considered striking up a healthy belief in God. It seemed like a good time for prayer, for taking stock of my soul and placing my life into more capable hands.

But how could I be sure that once I'd tossed my fate into the clouds, there'd be someone there to catch it? So I shelved the insanity and/or God question while I focused on the task at hand—the truck, not the wounded soldiers laying by the airstrip.

Take care, my friend. You're a fool. This is your battle, one you must win if you are to survive. This is one time that I will not help you.

We finished loading the truck and left for the fire support base.

I was disturbed at what I was becoming. My job—my very existence—was centered on the battalion. There was not supposed to be any time for self. Self was not supposed to interfere. We were genuine Government Issue and I, as an officer, was supposed to be immune to all but the mission. My mission was to keep the howitzers firing. Being human wasn't part of that mission.

—

War is simple. It's just numbers, just an equation. Supply and force make the difference between winning and losing a particular battle. At least that is what I had been taught.

Combat fire power = organic firepower (Those weapons built into the tanks and personnel carriers) **+ individual and crew-served weapons of the troops and supporting systems** (like aircraft and artillery).

If your side possesses more **combat fire power** than the other side, your side should win the battle.

In 1970, The 155 MM Self-Propelled Artillery Battalion was blessed with several pieces of first-generation equipment that were seen as significant advancements in our **combat fire power** as well as our ability to wield that fire power. Though this might seem like a beneficial advantage over other outdated models, the reality was that we were effectively field testing these brand-new weapons of war in the rather unfavorable conditions of Southeast Asia.

For the modern army, off-road capability was a key ingredient to every new piece of equipment. In the 1950s, military minds determined that the modern concept of land warfare would be waged through battlefield tanks supported by mounted infantry, so all the divisions would need to be adapted to fit this model.

Since tanks and troops were expected to go off-road in the course of the war, every piece of artillery was expected to go off-road as well. For the most part, this meant tracked vehicles, since wheeled vehicles tended to bog down in mud, snow and sand. Through the late 1950s and 1960s, the Army developed artillery equipment that was envisioned to respond to the most serious and logical threat to America, general war in the European theater.

But Vietnam was not located in Europe. The VC and NVA knew better. They used **bicycles**.

As the motor officer, it was my responsibility to ensure that the newest military models cutting their teeth in the forests and swamps of Vietnam would continue to function and live up to the expectations of the generals and politicians who had signed off on their production. Three particular pieces of machinery were sharp thorns in my side during my first combat experience.

First, let me introduce the **Field Artillery Digital Automatic Computer, (FADAC)** lovingly known as Freddy.

The Good Intention: Throughout history, the biggest criticism of artillery had been its inability to respond to a changing combat environment. This translates to slow response to combat emergencies. Freddy was meant to fix the slow artillery response time through quicker calculations required to aim and fire the howitzers.

The Reality: Freddy was a first-generation computer designed in the '50s. It was obsolete before it could even be fielded. Its two major flaws were revealed in Vietnam.

The first flaw was that its memory drum turned out to be sensitive to movement. So sensitive, in fact, that even the slightest jarring would cause it to lose memory and become useless. Specialized repairmen were required to fix the problem using equipment that was nonexistent at the battery level. Normally, Freddy would have to be airlifted back to a direct support maintenance company for repairs. This took days.

We quickly learned that Freddy could only be counted on to fail.

The second flaw with Freddy was its power source. It was a specialized, gas-powered generator that could only be used with Freddy. It was noisy and it ran hot. Because it needed a lot of air to keep cool, it could not be placed in a bunker to muffle the noise. The noise allowed Charlie to find Freddy very easily. Charlie knew that if he could knock out Freddy, he could screw up the howitzers.

Charlie hated the howitzers and therefore, by association, Charlie hated Freddy.

The troops did not like Freddy, either. They knew Charlie considered Freddy to be the most important target in the fire base, that Freddy was always located close to the Fire Direction Center. I didn't like Freddy because he caused me a lot of work since, as I mentioned, Vietnam was not what the designers had in mind when they built Freddy and his generator.

The only people who liked Freddy were colonels and generals. So if you think about it, there were definitely occasions when we liked Freddy, too. When the general's chopper landed, we really liked Freddy. And when it left, well, not so much.

Next up is the **M109A1, 155SP, Howitzer.**

The Good Intention: It was fully tracked, electrically and hydraulically operated, armor plated and capable of off-road operation while protecting its crew from small arms, nuclear blasts, and radiation. The hydraulics-powered rammer system was capable of ramming a shell into the tube's rifling at a much more consistent rate, which meant that a consistent amount of gas would escape from between the sides of the round and the rifled interior of the howitzer tub.

This was a good thing, since the lost energy would be more predictable and a more consistent velocity would be achieved for each artillery round. That meant consistent accuracy. Consistent accuracy meant the difference between dead enemy combatants and dead friendly combatants.

The Reality: In 1970's Vietnam, this expensive piece of metal was a 50-ton piece of junk. At first it had been pretty good at doing what it was supposed to do. But after some wear and tear, it became unreliable. When it came time to move, it might not move. When it came time to shoot, it might not shoot.

The only purpose for a M109A1, 155SP howitzer was to shoot a fire mission. It could be very exasperating when you tried to use a worn-out piece of junk to shoot a fire mission. The power pack inside the armor-plated shell of the howitzer was unserviceable at the operator level. We soon found out that its rubber hoses had a tendency to deteriorate in even the best of weather, let alone the wet-and-hot, dry-and-hot monsoon weather of Vietnam. When the hoses went, the howitzer stopped moving.

Another problem was that the hydraulic system was hard to keep in adjustment because the shock wave from firing the round sent a tremendous vibration throughout the howitzer, including the hydraulics-powered projectile rammer. It would break seals in the hydraulic system, tear up electrical wiring, and cause metal fatigue.

The power rammer was also unpredictable when it was new, and dangerous when it had some age. In my battalion, the only time we used a power rammer was when someone important was watching.

However, the most serious defect of the howitzer was its basic design. Like Freddy, it was too complicated for the operators to maintain in a combat environment. There were too many electrical systems, too many hydraulic systems, sighting systems that were too complicated, a power train impossible to maintain, and 50 tons of metal. The crew assigned to operate the equipment had neither sufficient training nor authority to fix the real problems. And when they were in combat, they had no time. The net result? This equipment didn't increase our effectiveness in combat in Vietnam. Instead, it worked the operators to death. Literally and figuratively.

Finally, let me introduce a companion piece to the **M109A1, the M548 ammo carrier**.

The Good Intention: The M48 ammo carrier was a tracked, thin-skinned vehicle designed to follow the M109 off road. It was an armored personnel carrier with a truck body. It was designed to be able to back up to the howitzer where the crew could offload the ammunition in direct support of the firing operation.

The Reality: It took great effort to service the most vulnerable parts, the rubber hoses. What was even worse was that the power train, the motor and transmission, was underpowered for the load it was supposed to carry off road. The power train was the same as that used in personnel carriers and, for that matter, the same motor used in Greyhound buses. It could not move the extremely heavy loads of ammunition required to support a howitzer without overworking the power train. After being in use only a short period of time, it simply broke down. The vibration of the tracks on the M548 caused the aluminum alloy body to come apart at the welded joints. The torsion bar suspension used in the track system broke down under the heavy strain of carrying ammunition. In my battalion, only two of the eighteen M548's assigned to us were operational, and they were both used to haul trash.

To further complicate my efforts to keep this equipment up and running, the pipeline of new equipment had been about cut off, so it became necessary to fix equipment rather than replace it.

This all might seem trivial, but there was a lot of junk in the hands of an army that had serious doubts about its mission in 1970. When you add to this army unit a leadership that wouldn't lead and troops who weren't sure they wanted to follow, you had the ingredients of a unit that was less than effective at best, and at worst, **bitter and counterproductive.**

Camp Price Battalion Compound, Long Binh/Bien Hoa May

May 1970

As time went by and I heard stories about my old unit, it became clear I was not the only fly in the ointment. This was oddly comforting.

The battalion ammo officer had been fragged as he left LTC Gray's hooch. The story was that they were after LTC Gray but got the ammo officer instead. It was a shame, because he was one of the few officers in the battalion who was liked by the enlisted men.

The Headquarters Battery 1st Sergeant was shot in the back with an M-79. It was at close range and the round had not armed yet, so the round just stuck in his back.

One day while I was getting a target and intel assignment at the group S-3 shop, I stumbled into CPT Mark Waters, the old Alpha Battery commander. He had been relieved of command and reassigned to the group staff as an Assistant S-3.

"Hey, Mark, how the hell are you?" I said. "I see that you have been delivered from the valley of the shadow, that snake pit of Hull. Fuck up and move up."

"John, long time no see. How's it hanging?"

"Not bad. I see they didn't get you with the AK-47. How did that come out, anyway?"

"Ring Knocker swept it under the rug," he said with a shrug. Ring Knocker referred, of course, to LTC Gray.

"How is Ring Knocker, anyway?"

"Hell of a deal," he answered. "Things are so bad in the battalion that they had to ship him out. They are investigating the place."

LTC Gray had actually been delivered to our battalion and had taken refuge in our battalion commander's hooch, but no one knew why. Now I knew.

No one had seen him leave the hooch since he arrived.

One evening, LTC Mullens, my new battalion commander, called me to his hooch. As I entered, I saw LTC Gray and I froze. I forgot to salute. He walked across the room and offered me his hand. I took it very weakly.

LTC Mullens said, "Captain, LTC Gray is interested in how you are doing in your new job."

"I am doing fine, Sir."

"I would like to give LTC Gray a memento of his stay with us. Do we have any VC sandals left?"

"Yes, sir."

"Would you mind getting a pair for LTC Gray?"

I wondered, bitterly, who would be the one to pay for them. I knew that asshole wouldn't. But I'd learned my lesson.

"I'll go get a pair and be right back, sir."

"Thank you, Captain."

I was steaming as I left the hooch. We were selling these VC sandals to battalion personnel as souvenirs, unique to our own unit membership. I used the money to buy supplies for our civic action projects. I shepherded my supply of sandals very closely because I didn't have many left and I needed the money desperately. I would rather give the sandals back to the VC then to give a pair to LTC Gray.

Then, as I walked to my office, I realized that if I gave him a pair of sandals, he would always remember me. I suddenly warmed to the idea.

I found the smallest pair in my supply, paid for them out of my own pocket, and took them back to the hooch. As I handed them to him, I knew my smile said all that needed to be said, but I did manage to say, "It's good to see you, sir."

"Same here, John. Keep your powder dry."

"Yes, Sir." I saluted both lieutenant colonels and left the hooch.

—

Being transferred to the new battalion gave me, for the first time since arriving in Vietnam, a chance to reflect on what was going on around me. The war began to confuse me. It was a series of dichotomies. I had come here to help these people, but to help them I had to try to kill them. I had come here in the name of freedom, but what I saw around me was not freedom. I had come here as a member of an army with a reputation of being masters at the art of war, yet the army I saw was no good. I was an officer and a gentleman, yet I had found hate in my heart. Worst of all, I was not the leader of my own destiny, but merely a passenger in a vehicle headed for God only knows.

While I hated it, every moment of it, I loved it. The excitement of the kill, the excitement of winning, was like a drug. Once ingested, never forgotten. It was all so very impersonal, and it became easy to develop an apathy to the fates of all those around me, including myself. But I did start to care for the South Vietnamese, seeing that these people were no different than my ancestors that won liberty two centuries prior. It was chilling to think of how the French had been so closely wrapped up in both our histories.

—

A good leader is an interesting animal. My first combat experience with a good leader came soon after my transfer.

He was my new Battalion Commander, LTC Mullens. He possessed a reputation for being a regular Army type with keen insight into human behavior. His unit was well led, properly equipped, and staffed to do the mission, resulting in a good combat record both on paper and in reality.

What I became most excited about was the unit's attempts to respond to the needs of the people it was trying to protect. Considerable time was spent trying to build and rebuild what had been ravaged by war. LTC Mullens saw that it was important to see beyond the perimeter to distinguish between friend and foe.

After I arrived, I reported to the battalion S-1, who delivered me to the battalion executive officer. The XO then delivered me to the battalion commander. As I'd assumed, my reputation had preceded me. I was received with a cool, suspicious glare.

"Captain Martin, you have been assigned to me as my new S-2. I expect that you will perform your duties accordingly. This battalion has an active civics actions program, of which I am very proud. We do not have an S-5, so you will be assigned the duties of the S-5 as additional duties. Do you think you will have any problems with the responsibilities of this assignment?"

"No, sir."

"Captain, I am aware of the problems you had at your last assignment. As far as I am concerned, they don't mean anything. As long as you do your job, you'll have no trouble from me."

"Thank you sir. I will do my best."

"I know you will. I want you to take a couple of days getting familiar with the job. I expect an intel briefing each morning before I fly and a second intel briefing at the evening staff meeting. You will give me the briefings. I can do without them for a couple of days, but you be prepared to start at 0600 day after tomorrow. The XO will give you the formats."

"Yes, sir."

"Stop by battery supply and draw a new issue of fatigues. I don't want my staff looking like derelicts. And go to the PX and get some personal articles. You smell like shit."

"Yes, sir."

"Do you have anything to say, Captain?"

"No, sir. Thank you sir."

"Dismissed."

I came to attention, saluted, and backed out of his office. It felt good to be back in the Army.

—

The second leader to repair my world was Staff Sargent Tran Van Son, an interpreter assigned to the battalion, which assigned him to me. SSG Son was a soldier in ARVN and he held the equivalent rank of a Staff Sargent in the American Army. Little did I know what this sawed-off guy with a swagger in his step and an ability to cuss in American better than most Americans would grow to be one of the most important people I would ever know.

Few of the officers of the battalion liked SSG Son because he was arrogant. He was a Vietnamese and he was proud of it. He would not cow down to an American, no matter how big or what rank. I liked SSG Son right off the bat.

A typical conversation with SSG Son went something like this:

"What the fuck, *Diwi*? Your god damned driver refuses to carry an M-16. No fucking way, numma fucking ten! I am not traveling all over South Vietnam with some Johnny-fucking-be-good!"

"Now, Sergeant Son, it isn't all that bad. We don't get shot at much."

"Fuck it, *Diwi*. Don't mean fucking nothing!"

It was kind of fun talking to SSG Son.

SSG Son was a North Vietnamese who had come south with his parents after the fall of the French colonial government. He was a small child at the time. The descriptions of the terror his family faced during their escape sent chills up and down my spine.

For him, as a little boy, it was a swim across a river at night, a swim into oblivion. He heard and remembered the sounds of death in the water behind him, but as a child, they didn't hold much meaning for him until he was older.

SSG Son carried his wanted poster in his wallet. It had been issued by the North Vietnamese. They considered him a war criminal and wanted him dead or alive. My SSG Son was just like a Vietnamese version of Billy the Kid. He had even been shot in the back. One time he pulled up his fatigue shirt and showed me the scar. He didn't talk about it, so I never asked.

SSG Son helped me coordinate my numerous activities both on and off the base. My new jobs as the battalion intelligence officer and the civic action officer were methods of gathering intelligence for the battalion. This important aspect of the art of war was in serious trouble within the Army during my stay in Vietnam. The Army's failure to prepare for the 1968 Tet Offensive was blamed on the intelligence community, and for good reason.

While we did rally and win the tactical battle, we lost the strategic battle. The net result was that after 1968, little faith was placed in the intelligence community by commanders. This was true, down to and including battalion-level intelligence.

This state of affairs provided an opportunity for me to experience my war in a new light. No one who mattered much cared what I did. But it is hard for me to overestimate how important this phase of my first war was to my eventual victory in my second, little war. The war that began once I returned home.

—

As the battalion intelligence officer, my only real function was to read the daily intelligence reports from the higher headquarters and brief LTC Mullens and the other staff on what they said. On the occasions when a firing battery would move, it was my responsibility to update and brief the battery officers as to the enemy situation in the general area of the new fire support base.

All this took maybe three or four hours a day. My involvement in the civic action program for the battalion took a significant amount of my time, during which SSG Son became an important fixture in my life. Twice a week, we conducted medical civic action programs in local villages we were trying to pacify or keep pacified.

The Med Caps were a joint venture with our sister ARVN battalion. We used these opportunities to train the ARVN medics as best we could. The goal of this program was that after we departed, the ARVN medics would know enough to continue the medical support to the villages.

We also had other civic action programs such as agriculture and construction in a joint effort with the local civilian and military governments.

During these trips, while interacting with his fellow countrymen, SSG Son opened my eyes to the fact that the Vietnamese were real people with real needs. He showed me that they bled and cried, loved and laughed, and found beauty in this world. Just like me. It was SSG Son who showed me what bravery really meant, and he gave me my first demonstration of true patriotism.

Like with many Vietnamese, there were those of us who had suspicions that SSG Son could be a spy in our camp. I never knew whether he was or he was not, nor did I care. SSG Son was motivated by a deeper truth than anything I had ever experienced. He saw the world clearly as fact of being, unlike me who only saw a series of dichotomies, of right versus wrong.

"Hell, *Diwi*, I know what I was; I can only assume that I will be something similar to what I am. Numma fucking one, no big deal."

"God damn it, SSG Son, we're trying to help you build a country here. Everywhere you look, it's all fucked up. If we don't tear it up, then Charlie does. How in the fuck can you stand there and tell me that it's no big deal?"

"Hey, *Diwi*, you got to learn to speak Vietnamese."

"You never answer any of my questions."

"Hey, *Diwi*, you got any gum I can give to these kids?"

"Shit, fine. Just make it go as far as you can."

II Field Force Headquarters, Long Bin Spring 1970

One morning during the monsoon, I met my Bird Dog at the Long Binh airfield. The Bird Dog was an observation airplane designed back in the 1950s to fly low and slow.

The time of day was early morning. It was dark, overcast and rainy. The monsoon was in full force and it had been so for quite a while. We took off, the pilot in the front seat and me in the back, and within a few feet of the ground, we passed into the clouds. We flew for what seemed to be several minutes, completely shrouded in a totally gray world.

Finally we broke out above the clouds. We were in a trough with great, pure white clouds towering all around us. Above us was the most beautiful deep blue sky. The sun was shining on the clouds so that they were the whitest of white. This was Vietnam for me, at once the grayest, dirtiest, most fearful and deadly, yet most beautiful place I had ever seen.

I was flying that morning because artillery had a nasty reputation of being unpredictable. The shells did not always land where they were intended. There were many reasons for this, such as human error, physical design, and the weather. Once a round leaves the tube of a howitzer, it kills whatever is unfortunate enough to be in the immediate vicinity of wherever it lands, whether friend, foe, or something in between.

Naturally this was concerning to those who had requested—and probably desperately needed—the support. In 1970, uncontrolled American artillery was as dangerous to American troops as was the enemy we were fighting.

To minimize the possibility of artillery accidents by a firing battery, it was the policy of the battalion operations officer that each battery should be checked each day by firing what was called a registration mission. Because of the jungle, the only effective way to do this was to fire the missions using an observer in an aircraft. This mission was being flown by members of the battalion operations section.

My other duties only required roughly three days of work, so I was looking for something else to do anyway and realized flying these missions would give me an opportunity to do my job better. I could familiarize myself with the areas of operation around the firing batteries, and I could gather intelligence that would directly affect our battalion.

And, if I was lucky, I might get the chance to kick the shit out of my most worthy opponents.

I volunteered to fly these missions every other day. This meant that on every second day I would fly from about 0700 in the morning to 1800 in the evening. It averaged about eight hours of flight time. During this time, I would adjust the registration missions as required by the particular firing batteries that I would be supporting that day. After I was finished with the registration missions, I would spend the rest of the flight time searching the country for targets of opportunity.

When combined with my more humanitarian endeavors during the rest of the week, this offered two intensely different views of the Vietnamese.

From above, it was impersonal as I watched the farmers toiling away in the fields with their children, ducking under the protective canopy of the jungle or watching my plane crawl along the sky. Perhaps deciding whether or not to try and shoot me down. But when I stood face to face with them, talking with them through SSGT Son, giving their children candy, and helping them rebuild the land I had helped destroy just the day before, it was very personal. They were my enemies, and they were my friends. I was there to help, and I was there to destroy. It was not a matter of choice. It was an adherence to the day's to-do list. It was the angle by which I met their eyes.

More sinister to me were the dichotomies within my own army. As the battalion security officer, it was my job to supervise the physical security of the battalion compound, which meant I became involved in the attempt to control the drugs and the race riots.

I supervised young men by day who I knew might try to kill me that night. An epidemic of drug trafficking throughout the country turned clean-cut young Americans into babbling idiots within a month, yet the battalion medical officer's—that was CPT Carr, or Doc for short—primary concern was the number of new VD cases from week to week. The typical nightly battalion staff briefing for the commander included a report from Doc Carr about the VD rate, was it going up or down. He never discussed the level of drug use in the battalion. After all, if the higher-ups insisted that there wasn't a drug problem in Vietnam, then there must not have been a drug problem in Vietnam.

As the battalion security officer, I figured it was important for my security force to be somewhat coherent. I assumed a drugged-up soldier was not part of the security solution. On the other hand, a soldier with VD could fight, so I was not all that interested in the VD rate.

Racial tension was also a dark secret of the Vietnam-era Army in 1970. The Black Panthers had organized to the point that there was an army within the army. Doc Carr and I worked closely with the battalion medical civic action projects (Med-Caps). The grapevine had it that Doc Carr was also in charge of one of those Black Panther units. He was a dedicated physician and I respected him. Though we were never really friends, he would come through for me in a few very crucial moments in the months to come.

Our Battalion Compound
Late Spring 1970

During the battalion commander's morning briefing, LTC Mullens told me there was a dog running lose in the battalion area and that I was to find it, kill it and get its head to Doc Carr. He was concerned the dog might have rabies. I decided to get this done before I left for the Sugar Mill to complete a solation payment.

The dog had taken refuge under the trailer because she felt safe there. I'd just shot her with my war trophy, a Chi-com 54 service pistol used by the NVA. I had walked up to her and placed the muzzle of the pistol behind her shoulder and fired. The ammunition for the Chi-com 54 was steel jacketed, so at close range, it would pass right through a body and cause very little damage. The pistol was poorly made with a very sloppy action. I had to be standing right next to my target to hit it. It was mostly just good for being loud and scary when fired.

The bullet passed through her body without causing any serious damage. She ran to a water trailer and took refuge.

So I asked an E-4 standing next to me if he would shoot her with his M-16. He refused.

I took his rifle, went over to the trailer, and knelt down. As she looked up at me, terrified, I placed the muzzle of the rifle under her chin and fired a round. Her body jerked at the impact. She hadn't expected me to pull the trigger. The look in her eyes had been one of acceptance. She had thought I was going to comfort her.

She just continued to look at me with those big brown eyes. The bullet entered just above the breastbone at the neck and traveled the length of her body. I had never seen what an M-16 bullet would do at close range. I expected it to blow a large gaping hole where it exited her. As the bullet passed along its journey through her body, I was surprised at how long it took for me to see any evidence that anything was happening. The bullet must have been causing havoc inside her, but she just continued to look at me with those big brown eyes. They did not even blink.

What if it doesn't kill her, my friend? What will you do next?

Finally, the bullet spurted out just above her left hip, leaving just a small hole and a little blood.

She never even blinked. Slowly, so very slowly, her head lay to one side and her eyes began to see a reality of a greater dimension than mine. In those few seconds that seemed like hours, her life imprinted on my soul.

I don't know how long I remained there on my knees by the trailer, but finally I realized that I had to do something. I got up and walked back to the E-4 and handed back the rifle. I told him to put the body in the back of the three-quarter-ton truck and take it to the aid station. Doc Carr wanted to examine the body to see if she had rabies. I glanced into the soldier's eyes and I felt his hatred for me.

As I turned to find SSG Son, my stomach was tied up in knots and I thought I was going to vomit. I had done what I had been ordered to do, which was not that different than when, several nights before, I had ordered an artillery strike that killed 20 people in a VC rest area. Why did I feel pride in killing humans from afar, but damned for killing a dog suspected of having rabies? Somehow, this was different.

I had never killed animals at such close range, nor had I ever killed anything that had trusted me. Those eyes burned a deep impression in my mind.

I understood why she had to be killed, so that Doc Carr could cut off her head and test her for rabies.

The dog had belonged to an enlisted man who had bought the farm on a convoy several weeks earlier. The dog represented memories that several friends were not willing to give up. But not one of them was willing to take care of the dog. The result was that the dog became a health hazard to all of us, including these so-called friends. It hurt to kill the dog, but it made me mad that it was necessary to do so in the first place.

How do you kill a dog with honor?

But it was time to get going. It was a long way to the Sugar Mill and I sure didn't want to get caught out there after dark.

As I sat in the Jeep, I kept going over and over in my mind how she had looked at me, looked through me and finally looked at me not at all. I caught myself wringing my hands and folded my arms over my chest with my hands closed into fists. I didn't want anyone, especially SSG Son, to see that I was nervous. There was a feeling in my body, in my arms and in my chest. It was kind of like pain, but not really. It was like tingling, but not really, it was like excitement, but not really. I wanted to cry.

The voice;

I am worried about this guy, I don't think he is getting it.

Vietnam is a very strange place, one minute the pucker factor is so high that you feel like you are going to explode. The next moment the peace and beauty of the country is overpowering. The real trouble with Vietnam is that you never know when to expect the terror or the peace. You can't predict the present, little alone, the future. That is very unsettling.

For example, when the sun went down back in the world, you could predict a safe and secure night. Here, if it wasn't incoming, it could be sappers, rats, Dopey Doug's fragging officers, race riots, or Black Panther Pete's bullet in the back. This shortened our ability to predict the future.

In the World, you see a pretty girl and you try to predict the possibility of a long and positive relationship. Here, you wonder if she will be the one to cut your throat or give you a dose of incurable clap so you're sent off to that island in the Pacific where the Army stashed all the soldiers too diseased to send back to the World.

In the World, if all else fails, you have a family. Here, you are alone. In the World, you have a future. In Vietnam, if you have a future, it is never.

The voice, with a little humor;

It is enough to drive a man dinky dau, isn't it?

The only logical way to deal with such an environment was to develop a safe method of withdrawal. A way that can be turned on and off at will and still retains enough control to allow the user a method of overriding utopia. In Vietnam, in 1970, there were several popular ways to do this. There were drugs, alcohol, girlfriends, boyfriends, and a bullet –the final step over the line. They all had risks, and from my point of view, the risks were unacceptable. Worse, they all had obligations, and I certainly didn't need more obligations.

As the Jeep progressed toward the Sugar Mill, for a moment the fog began to lift. I could feel deep within that I wanted to do right.

But I was not sure that I was in control.

Long Bien, Vietnam
Late spring 1970: a day earlier

During the Cambodian invasion, we provided support for Task Force 333, which invaded along the border in the area known as the Parrot's Beak. Our contribution to that effort was one firing battery of six 155 MM towed howitzers. The towed howitzer was air mobile, meaning a helicopter could pick it up and quickly move it to locations inaccessible to wheeled or tracked vehicles.

To expedite the move, the firing battery would travel by road as far as possible. Then the helicopters would rendezvous with the convoy at a predetermined location and airlift the battery to the new fire support base. There were long periods of waiting for the convoy as the helicopters shuttled back and forth between the rendezvous and the new fire support base. The wait was boring.

There was a natural attraction between the local inhabitants, especially the kids, and the soldiers and their equipment. Everybody wanted to play. The soldiers would give out candy and buy cold drinks. The kids wanted to explore the military equipment and play with the soldiers. They would climb all over the equipment, playing whatever game they found interesting at the time.

During one of these games, a little girl fell off a five-ton truck and struck her head on the ground. She received a serious subdural head injury. A medevac helicopter was called, and she and her mother were air evacuated to the hospital at Cu Chi and then to the hospital at Long Bien for surgery.

I learned about the accident at that evening's staff meeting LTC Mullens instructed me to investigate it and to make a solation payment if I determined one was warranted. A solation payment was an act of condolence when our non-combat actions caused injury or property damage. The payment was a gift intended to show our concern and remorse.

Since the little girl was in the Long Bien hospital, I asked Doc Carr if he wouldn't mind coming with me to visit her and her mother. I knew he had several friends there and that he knew the hospital well. He said he needed to scrounge some medicine for the MedCaps (Medical Civil Action Projects) anyway, so he agreed to go.

We entered the hospital at the north end of the neurological ward, through an exit door at the end of the bay. I closed the door and turned around. It took a couple of seconds for my eyes to adjust from the harsh outside daylight to the soft, subdued light of the bay. As it came into focus, the bay seemed to stretch to infinity. There were two rows of beds, one row on each side of me.

The beds were full of naked American soldiers. Each soldier had multiple tubes stabbing into his skin and a massive bandage covering his head injury. Neither the doctor nor I moved; we just stood and stared. I didn't know how Doc Carr reacted, but I was stunned. We were in a neurological ward in the middle of a war zone and I was stunned because the ward was full of wounded soldiers. Didn't make a lot of sense.

One of the soldiers turned his head toward us and moved his arm, trying to pull the sheet up over his midsection. The sheet wouldn't move because of the tubes. It had to hurt, I thought, trying to move with those tubes sticking out all over the place. The soldier's eyes were open and he was looking at me.

I felt a tingling sensation.

What do you see, asshole? His eyes are empty. He is empty. Do you think they are draining his soul with all those tubes? What does one do with a spare soul?

Needless to say, I was not prepared for this. I mumbled: "I don't know."

Doc Carr looked at me and we quickly moved through the ward to find the medical staff. I looked straight ahead. I didn't want to see any more eyes.

I was struggling. I had forgotten the little girl's name. I couldn't make my mind work.

Don't you think it is very strange that you can't even remember that little girl's name?

When we found the doctor who had worked on the little girl, he told us she had died during surgery. He did not know where the mother was, so I assumed she had returned to their village.

A solation payment was warranted.

The regulation on solation payments established guidelines as to what was an appropriate payment for the death of a man, woman or child. A male child was worth about fifty dollars, compared to around a hundred bucks for a man. Girls were worth less than boys.

The process for making the payment was somewhat ceremonial. It was to be given to the parents in the privacy of their home and in a manner that ensured the family would not discover its value until after we left. The gift was to be a private affair for the grieving family.

In Vietnam, a gift of solace should be graciously received and a gift of appreciation should be graciously offered when the intent of the party responsible for the act and the intent of the aggrieved party are to mend the grievance. The regulation advised that we should be prepared for the family to respond to our offer of solace by offering a meal as their gift of appreciation. We were to graciously receive any gift the family might offer. In the process of the giving and the receiving the hurt would begin to mend. That was how it was supposed to happen—so said the regulation. But it wasn't quite that simple.

Vietnam/Cambodia border
The day of the dog: late spring 1970

The time had come to meet with the little girl's family and make the solation payment. After I had brief LTC Mullens on the death of the little girl and my decision to make the solation payment I met with Doc Carr at the aid station to work out the details for getting the dog's head ready for testing. We got into a pissing contest as to who should detach the head. Doc Carr thought I should do it. I thought he should do it.

I told Doc Carr that I would kill the dog because LTC Mullens ordered me to do so. But, if he wanted the head cut off he could do that himself. I would not touch the dog. Then I commented that the reason we had aid stations was to deal with blood and dead things. I told him to do what he was being paid to do.

We glared at each other for a couple of seconds and then I left to find and kill the dog.

After I killed the dog, I had the carcass loaded into a three-quarter ton truck to be delivered to the aid station. That was the end of that, or so I thought.

Then I rounded up SSG Son, my driver, and my Jeep, and we left for the Sugar Mill.

The country we were traveling through was a little piece of Vietnam that had not been devastated by the war. I was upset about having had to shoot the dog, but the beauty of the scenery calmed me and I eventually forgot about the dog. It always surprised me to find that there really was more to Vietnam than the depressing places like Tay Ninh, Cu Chi, Long Binh, Zuan Loc and Phu Loi. Their natural beauty had been Americanized and damaged by the war. But out here it was still real, out of character for the war.

I wanted to try out a new camera I had purchased at the Post Exchange in Long Binh. This trip would be my first opportunity to do so.

I had not given much thought to what would occur when I finally met the little girl's family. I truly felt for them. I had a little girl. I wondered what it would be like to be the father who is grieving for his dead daughter. A father who would receive a visit from a strange soldier who could not even speak his language, and who was not be trusted, a soldier who might be my enemy. What would I do if things were reversed for me? Would I let him into my home? Would I talk to him?

Then an unsettling thought crossed my mind. What if this father is a VC?

As I sat in the passenger seat of the Jeep and looked out over rural Vietnam and thought about what I was about to do, I forgot about the dog. It didn't help that my driver, a conscientious objector named SPC 4 Clark, refused to carry a firearm. Instead, he carried a Bible, which he read every chance he got. He did what he was told to do, did it well and did it on time, so long as what he did caused no harm. His manner was unobtrusive. And that was OK with me. I figured SSG Son and I could handle anything that might come up.

SPC 4 Clark was a funny sort, He had been assigned to me about a week earlier. He played the part of conscientious objector correctly. He convinced me he was no fake. I guess he was assigned to me because they couldn't find anything else for him to do. In my first meeting with him, I told him that I didn't care if he didn't want to carry a weapon, but that I wouldn't protect him since he wouldn't protect me. He said he could live with those rules. I didn't tell him that if things got bad I would do whatever it took to protect him. He didn't need to know that. His safety was my responsibility.

Our first stop on our way to meet with the family was at the Sugar Mill, which was the headquarters of the local ARVN security force and the American liaison team. It was collocated with the appropriate provincial governmental authority.

There is a right way to find people in Vietnam. It is not a matter of looking up their address in the phone book. It is necessary to get permission from several people in authority. First, I had to talk to the district chief, then the Military Assistance Command, Vietnam (MACV) advisers, and then to the village chief. With their permission, I could then go to the girl's home and give the solation gift to her parents. Getting the arrangements all worked out took a lot of time, and while we were doing so, the word of our coming spread throughout the local Vietnamese community.

Our last stop for permission was the village chief. As SSG Son and I entered the village chief's office, I noticed that he had a picture of Ho Chi Minh on the wall. The picture put me a little off balance.

The meeting was brisk and formal. While I did not understand the conversation between the village chief and SSG Son, I could tell by the tense tone of the conversation that we were not being welcomed with open arms. The village chief would look at me, his eyes moving from my head to my boots and then back up to my eyes, with an expression that let me know he did not like what he was looking at. Finally, he shook his head in an affirmative and dismissive way and waved his hand to tell us we were dismissed.

We excused ourselves and left. On the way to the Jeep, I asked SSG Son if he had noticed the picture of Uncle Ho on the wall. He responded that he had noticed it.

"Does that bother you any?"

"No, Ho Chi Minh is a national hero to most Vietnamese."

"Why?"

"He helped liberate Vietnam from the French."

I thought about what he said.

As we got into the Jeep, I turned to him and said, "Let me see if I have this straight. Here we are, almost in Cambodia. Alone, I might add. We just got finished talking to a guy who worships Ho Chi Minh and who runs a village which also worships Ho Chi Minh. We are now going into this village to find a mother and father and tell them we are sorry we killed their daughter. On top of that, our driver refuses to carry a weapon. And you say it doesn't bother you?"

"No sweat, *Diwi*."

"What do you mean, 'no sweat, *Diwi*? Number fucking ten."

Clark was staring at me. His eyes looked like eggs with a dark spot in the middle.

"Clark," I said, "do you have your Bible with you?"

"Yes, sir."

"Then I suggest you do some serious fucking praying."

I stood up in the Jeep, put my helmet on backward, did my best impression of Eric Von Zipper the Third from *Bikini Beach,* and yelled, "Charge!"

SSG Son laughed, and away we went.

Clark didn't say a word.

We drove to a row of small homes along a canal. The canal was on the left and there were rice paddies on the left of the canal. The homes were accessible by a foot path between the canal and the homes. The foot path was wide enough to drive the Jeep down, however, it was not wide enough to turn the Jeep around.

It was a perfect place for an ambush.

SSG Son and I left Clark behind with the Jeep in a big open area by the canal where he'd have plenty of room to maneuver the Jeep. If the necessity arose, we would be able to make a quick getaway.

SSG Son and I had to walk some distance before we could even get close to the girl's home. As we approached the row of houses we were looking for, we spotted a large group of men, women and children gathered in front. They were quiet and subdued. It was a good sign that there were women and children in the group; the probability of an ambush was minimal.

However, nobody would look us in the eye. This was not a good sign. Normally when I joined a group of Vietnamese, it felt like a sea of people closing in around me. I was a curiosity, something to be touched, smelled, and wondered about.

But not today.

As we moved through the crowd, they stepped back to let us pass. There were no comments with responding laughs, no touching, and the children were physically restrained outside my reach. I could not make eye contact with anyone.

We walked up to the gate for the little yard in front of the house. A woman stood behind the gate, but she was not the girl's mother. She turned out to be the girl's aunt. The parents weren't home. I waited as Son explained why we were there. When he tried to give the envelope, the gift, to her, she refused to take it.

A cold chill shot down my back.

This is it, I thought. *I'm going to die.*

I noticed SSG Son was sweating. I'd never seen him sweat before. I could tell he didn't know what to do. So I ordered him to force her to take the envelope and open it. He turned and looked at me as if to say that is not the way it is supposed to work. I nodded at him as if to say I know, but do it anyway. He turned back to the aunt and growled something in Vietnamese.

He grabbed her hand and forced the envelope into it, continuing to growl at her in Vietnamese. The aunt looked down at the envelope in her hand. Then, with us still standing there, she opened it.

When she saw the money in the envelope, she stopped being resistive. A look of surprise, or maybe disbelief or confusion came into her face. She looked at the money, and then at Son, and then at me. Whatever she'd been expecting, this wasn't it. She took the money out of the envelope and spread it out like a fan to show the crowd.

And she smiled.

Instantly, the atmosphere changed. She began to talk to the people around her and they moved to see the money in her hand. They would look at the money and then at Son and me. Here and there I saw a smile. A young boy broke loose from his dad, walked up to me, and touched my uniform. I smiled at the boy and he smiled back at me. People began to talk, and the tension began to dissipate.

In his best GI English, SSG Son turned to me and said *"Diwi*, let's get the fuck out of here."

I nodded my head in agreement.

As we disengaged and moved through the crowd back down the path toward the Jeep, an old man leaned close to SSG Son and whispered in an urgent tone into his ear. Without hesitation, SSG Son took off like a shot running away from the crowd. I sprinted after him, but he was moving too fast for me catch up. Clark saw us running for the Jeep. He was in the driver's seat and had already started the Jeep. He spun it around toward the exit road. When we jumped in. I yelled, "Don't let no grass grow under this fucking Jeep."

Clark didn't. When we left the village, he had the Jeep engine whining and the transmission in fourth gear.

A little later, as we began to settle down, I asked SSG Son what the old man had said to him.

"He said the VC were there. They had told the people that we were going to try to trick them, that there would be no money in the envelope. The VC told them we would keep the money for ourselves. The old man said the VC promised the people that they would make things right."

I thought about that for a second or two.

"I think that was a hundred bucks well spent, don't you?" I asked.

"You got that right, *Diwi*."

We had not driven far from the village when Son yelled, "Stop! back up!"

Clark slammed on the brakes, jammed the Jeep's transmission into reverse, and gunned the engine. He probably thought we were in an ambush or were about to hit a mine.

"Look at that," Son said and he pointed to paddy dike on the left side of the road.

An NVA soldier lay by the road on a stretcher. The stretcher was leaning against the rice paddy dike. The soldier's North Vietnam Regular Army uniform looked new. I thought that rather strange.

I noticed that he was an officer. He was big for a Vietnamese. His hands were bound above his head with commo wire and the commo wire was tied to the wooden handles on the top of the stretcher. His legs were spread and his feet were bound to the wooden handles on the bottom of the stretcher with commo wire. His body and head were slouched down on the stretcher.

His shirt was open. The blood on his chest and on his neck was black. What was left of his eyeballs hung down from closed eyelids. His throat had been cut. He must have been dead for a while, because he was turning black in the sun.

His torso had been wrapped in chicken wire and, where it touched his flesh, the wire cut into him and disappeared into his body leaving a series of cuts and dried black blood on his skin. The chicken wire would come out again when there was clothing covering the skin. I had heard of this type of torture. It was called the death of a thousand cuts.

"Hey, *Diwi*, take some pictures."

"No, Tran," I said. "I don't want to remember Vietnam this way."

I began to feel sick. It took all my effort not to vomit. When I regained my composure, I looked at Son. He was sitting in the back seat of the Jeep, staring at the body. I wondered what he was seeing and thinking and feeling. I remembered the price on his head, and wondered if he saw this as his own fate.

I didn't look at Clark as I ordered him to drive on.

Mine eyes hath seen the glory of the coming of the Lord, I thought.

Don't forget it, asshole. Got a question for you, dimple brain. Charlie there, he saw something not too long ago. What do you think it was? Was it glory?

Traveling back along the dirt road toward Cu Chi that day was a very quiet affair. There was no talk about Vietnam, home, the refrigerator that SGT Son wanted me to get for his parents, or the war. The three of us were busy dealing with our own private thoughts.

We arrived at the main gate of Cu Chi at the same time as a convoy coming from the northwest, so we had to sit and wait until it had passed. On the right shoulder of the road beside the Jeep sat a Vietnamese *mama san* and her daughter. I noticed the girl had no legs.

As we sat in the idling Jeep, the mother stood up and walked to the side of the Jeep and said something to SSG Son. He listened to her then turned to me:

"*Diwi*, she wants to know if we will give her and her daughter a ride to their hamlet. It is about three klicks down the road."

There was no reason not to give them a ride, so I agreed to her request.

The mother went back to her daughter, picked her up, and returned to the Jeep.

I got out and tilted the passenger seat forward so they could get in and ride on the backseat with Son. I reached out to take her daughter into my arms while she climbed into the back of the Jeep. The mother handed her daughter over to me with a slight smile on her face. The little girl was so light, so small, and so fragile that I was afraid that I would drop her or somehow break her. I was relieved to give her back to her mother.

As we started out again, I asked Son to find out how the girl had lost her legs. The mother told him an American Army truck had run over her. Her legs had been amputated at the hip, and now the American doctors at Cu Chi were trying to rebuild her lower body. It wasn't an accident, she said. The driver had run the girl down intentionally.

I felt a tear cut a path through the dust on my face.

This was not an accident of war. This was a crime of vicious and wanton behavior by some American soldier. There was nothing I could do or say in response to what I was hearing.

This day was difficult. It was about Americans killing or maiming Vietnamese children and I was in the middle of it. I couldn't stop it, I couldn't help it, I couldn't make things better. I was helpless.

I sat in the Jeep and looked down the road.

This time, the taunting voice in my head was my own;

Listen up, dickhead. This is glory. This is fun. This is war. You volunteered for this shit. This is the society you wanted to be a part of.

Open your eyes and see how their glory shines.

You came over here thinking you were going to be a hero, to do as those before you have done, and save the world from the great wrongs others created. You came over here thinking that your country is a great nation and that you are going to fight and die in a blaze of glory.

Wrong, asshole. You are not going to die. You are going to live, you are going to see, and you are going to remember. America sucks, you suck, and you are wrong. Everything you are doing is just a big fucking joke. You are a big fucking joke. Shit, man. Do you think that they give a damn about what you do? They don't give a shit that you are ashamed about what happened to this gook.

Don't mean nothing.

I knew that I would never be able to forget that little girl. She represented everything that was wrong with me, my world, and my war. She was that which was right, and we destroyed her. She was that which was good, and we destroyed her. She was that which was beautiful, and we destroyed her.

As we drove down the road, that little girl laid open all that was wrong in my heart. She introduced me to the sin of arrogance—my sin, my arrogance. I deserved it.

When we reached their hamlet and their home and I started to get out of the Jeep, I fought to control my emotions. I exited the Jeep, lifted the passenger seat forward, and reached out to take the little girl from her mother. She reached out to me and as she came into my arms, she studied me. I held her tight and with one hand cradled her face into my neck. We waited like that while her mother got out of the Jeep.

Before I gave her back to her mother, I looked into her eyes. "Tell her that I am ashamed that one of my countrymen did this to her," I said to Son. "Tell her that if I could, I would give her my legs. Tell her that I wish that there was some way I could make her right again."

After Son told her what I'd said, she reached up and touched a fingertip to my face to catch a falling tear. In her eyes I saw the power of her will and her capacity to love.

I felt as if I were in the presence of God.

When we were back on the road, Son reached up and gave my shoulder a gentle tug. During the final part of our trip back to the battalion, I felt a little better about this horrible day.

We got back just in time for the evening briefing. As the S-2, I was the first up. I told LTC Mullens that I had made the solation payment. I explained that I had just returned and had not had time to prepare for a briefing.

That was the end of that.

I listened as the S-3 described the operation at the Sugar Mill, and the S-4 give the status of the airlift. I listened as Doc Carr explained about the dog. It hadn't had rabies after all, he said.

Well, shit, I thought. I had killed a dog for no reason. *Wonderful.*

As I half-listened to LTC Mullens explain how the Cambodian operation was going, my mind wandered back to that little girl's eyes and her love of life, and what was probably the terror of torture and death, and the mistrust of so many, and the bone-chilling fear I had felt in that village. I was glad that I was safe in the back of the room where no one could see me.

You have been touched by a greater will. Much of your true worth was not being used; much of your life has been wasted. It is time now for you to start anew. It is time for you to open your eyes. Start again from this new beginning. See your world for what it is. It is your choice to experience good or bad. My friend, in the final moment of truth, it will be as you decide it should be.

I chose to cry; I chose to love. I began to realize that this war was like something else of major importance in my life—winters in Wyoming. When winter came, it was all-encompassing. It took total control. It was absolute. It purified all that it touched, and it made way for new beginnings. I knew that for all I had seen and all I had experienced, had it not been for this war, I could never hope to be whole.

At least now I had a chance at life.

Later that evening, I stopped into the officer's club for a beer and played a couple games of pool with Chief Warrant Officer Parker, the battalion maintenance warrant officer. Parker ran the battalion maintenance section. We crossed paths because I had experience as a battalion motor officer. Because he served as motor officer in addition to his normal job, I volunteered to help him where I could. We both liked pool, so every night he and I would play for a while at the officers' club.

The TV over the bar had the news on, but no one was watching. I heard them saying something about the National Guard and Kent State. Then I heard the phrase, "several dead and injured."

I stopped playing pool and turned toward the TV. There was this sweet young round-eye (an American female college student) on her knees crying and holding some guy who had just been shot. After the segment had ended, I turned to Parker. "Chief, those assholes are trying to get us killed."

"Sure looks like it, don't it?"

"I wonder what would have happened if that had been a platoon from the 25th Division or the 11 ACR instead of a National Guard unit," I said. "Give me an M-60 and I would have given them something to really cry about."

"I hear ya."

South Vietnam
Spring 1970

As the days passed, I started to develop a routine. Every other day I would fly in a Bird Dog as an air observer, familiarizing myself with the terrain around our fire support bases. When I returned from flying, I would spend a few hours in the evenings getting up to date on the intelligence reports from 23rd Artillery Group and II Field Force. I became more in tune with the intelligence side of my small corner of the war.

On the days I did not fly, I spent most of my time with the civic actions projects. Most of our projects were medical in nature, but we did have a few that were not, including an agriculture education project. We worked with selected prisoners of war to teach them how to raise chickens with feathers.

One of the first impressions I had of Vietnam is that Vietnamese chickens didn't have feathers. They really were an interesting creation. A featherless chicken looks, somehow, indecent. It became clear that I wasn't the only American who thought so, and in typical American style, someone decided we should do something about it. Never mind that the Vietnamese chicken had been featherless for generations, and never mind that the Vietnamese liked their featherless chickens, and never mind that featherless chickens cooked up into some rather tasty morsels. Their chickens were absolutely, positively indecent, and we were going to do something about it.

It so happened that we had some farmers in our unit, so the chicken project became mine. We decided to teach prisoners of war how to raise chickens without the disease that caused their feathers to fall out. In return, the POW's gave us VC sandals that we sold to buy supplies for our medical civic action projects. The real goal of the chicken project, however, was to make the POWs like us. That's what we hoped for. I never found out whether POW's began to think of us as good guys or not. Didn't matter to me; I got my sandals.

SSG Son and I spent as many of those days as we could away from the battalion. It was easy for us to get away because I had legitimate business in many areas in and around Bien Hoa. I had a special pass that gave me full access to most of the off-limits areas. This meant that Son and I could go places where no one else in the battalion could go, including my boss, LTC Mullens.

It became a common occurrence that on each trip we would recon something about the real Vietnam. I was particularly interested in the different types of art, architecture, and religious symbols, so Son made it a point to educate me as best he could.

As the days were full of learning about Vietnam, the good and bad, the nights were filled with learning about the US Army.

Intelligence information gathered in report form is very perishable. If the information is not acted on promptly, the targets disappear. It did not take me long to figure out that there was a problem with the intelligence sent by higher headquarters about enemy troop movements. It was obsolete. By the time II Field Force had staffed it and passed it to 23rd Artillery Group, who then messed with it until they were satisfied, republished it and passed it to us, it was days old. Except for general information purposes, it was useless. It had no target value at all.

Each morning, I would go to the operations center at II Field Force to get the latest intel update. There was never anything fresh enough for me to react to other than to note for possible observation the next time I flew in the area. These same reports were hot stuff for the II Field Force staff, but as far as I was concerned, at battalion level, they gave me nothing to shoot at.

It was frustrating to try to explain to the starched uniforms that the reason we got no body count was that we got no information we could use. I would try to explain that it was frustrating as hell for these howitzer crews to shoot fire missions all night long, fill sand bags all day long, and never hear any feedback indicating that what they did had any effect one way or the other. After a while, the howitzer crews just stopped caring.

It became clear that my higher headquarters was not going to provide me with the information I felt I needed to do my job correctly. They knew who had the real hot stuff. But they would not let me know who it was. I decided to try to find the intelligence sources myself.

I stumbled onto one of them by accident.

I was given an observation assignment by the Group S-3 that required me to coordinate the mission with a sister battalion. I had gone to the operation's section of the sister battalion to do the final coordination and was talking to the intelligence sergeant when I noticed his map was set up differently than mine. It had more information on it than either 23rd Group's or II Field Force's. A lot more information, in fact.

The intel sergeant told me that he was not sure who sent in the intelligence, but it seemed to be real hot stuff. At one time, this battalion had our mission as Long Bien Special Zone Control; but, they no longer had any need for the information because their firing batteries were no longer in the area. Nothing was ever done with the information.

The sergeant was able to give me a telephone number, as a point of contact, but no name. Our mission as Long Bien Special Zone Control meant we were in contact with all artillery units in the area, and we kept up on their fire missions so we could direct aircraft around dangerous areas. Because we had contact with everyone, we were in a perfect position to call immediate artillery fire support. The only problem was that we had no targets to shoot.

I called the number the sergeant gave me. An American answered. I explained who I was and what I wanted. He put me on hold, and after a minute or so, another American voice came on asking for me to explain again what I wanted. I told him that I wanted up-to-date intelligence information so I could engage targets with artillery firepower.

"Explain to me what you mean by 'engage targets with artillery firepower'," he said.

"Well, if I had good intelligence information, I could arrange to have artillery rounds hitting the target area in minutes of the time I got the information."

"How much artillery?"

"Depends on the target and the target location."

"You mean that, if I gave you a target to shoot with artillery, you could do it at a moment's notice?"

"Yes, Sir. Well, not quite a moment's notice, but fast enough."

"I'd like to meet with you, Captain."

"I'd like to meet with you too, sir. Give me your location and I will come over to arrange something."

"That's not possible, Captain. But I can meet you someplace. Would it be possible for me to come to you? Would it be possible for us to see your operation?"

"Yes, sir. How about ten hundred hours, tomorrow morning?"

"Where do we go?"

"We're on the road between Long Bien and Bien Hoa. The road that goes by Widow's Village. Do you know which one I'm talking about?"

"Yeah."

"We're in Camp Price just west of Widow's Village. Stop at the main gate and have the guard call me at the Tactical Operations Center (TOC) and I'll come and get you."

"OK, I'll see you in the morning."

The next morning, the guard at the front gate called to inform me that there were some people wanting to get in the compound. I told him to hold them until I got there.

Their civilian clothes and their gray Jeep told me that I had struck the jackpot. These guys were probably CIA, CID or spooks of some unknown heritage. It didn't take me long to realize they were the guys who had the spies in the field with the enemy troops.

My chances to shoot good targets were not going to get any better than this.

Since I did not know them and I did not have any formal clearance for them to view classified information, I checked their IDs. They had the proper IDs. If they were who I thought they were, they would understand my reluctance to expose them to any classified information. I took the two guys to an area in the operations center we used for meetings, where we had no classified documents or maps exposed. I explained in detail how we could call artillery on any target in the general areas of Long Bien, Tay Ninh, and Xuan Loc. All I needed was some information that would make the fire mission worth the effort. I explained that the system we used to get intel from higher headquarters was not working because of the delay inherent in the system. I explained that what I needed was up-to-date information.

The guys in gray listened for a few minutes, then asked if I would demonstrate how we shot targets. They gave me a target they wanted shot. It was a bridge over a small creek where Charlie was moving supplies. They indicated the bridge was the primary target and they wanted it destroyed, but added that we might get troops and equipment, too.

I wanted to impress them with the extent of our capabilities, so I set up several firing batteries to complete the mission. I used eight-inch howitzers and 175 millimeter guns from one fire support base to get the heavy boomers, a 155 millimeter battery for accuracy, and a 105 millimeter battery for additional effect. I arranged what we called a "time on target," or TOT, meaning all the rounds from all the units arrived at the same time.

The reason was surprise. If there were troops in the area, they would not be warned of incoming before it actually occurred. I mixed the type of artillery rounds to be used so that there were anti-personnel rounds going off at the same time as the rounds designed to destroy equipment and structures. I used the eight-inch to go after the bridge, and the 155s and 105s to go after the people.

I explained that there would be no guarantee we'd get the bridge. "Since we don't have eyes on the ground, we can't adjust the artillery," I told them. "The chances of us hitting the bridge are slim. But if Charlie is there, we'll ruin their day."

We shot the mission. One of the guys in gray wanted to know how we had done. I explained that the mission was unobserved, so until we could get eyes on the ground, we would not know the results. Since it was their mission and they wanted to know the results, I suggested they get a chopper and go see for themselves how we'd done. If the bridge wasn't damaged, I'd go out there in my Bird Dog and destroy it the next time I had a mission in that area. I said if they needed more artillery support while they were out there, they could give me a call and I would arrange another mission. I gave them my call sign and radio frequency so they could give me a damage estimate, and told them it was important for me to be able to tell the firing batteries how well they did.

The guys in gray left and got a chopper.

When they radioed me later, they sounded ecstatic. The results were better than they had expected. All of a sudden they had a way of getting at the targets when all else failed. They had a way that was accurate and not subject to the weather or human frailties. That night they called me with another mission and I agreed to fire it, no questions asked. They couldn't believe it. Finally they could get an aggressive response on a target of interest.

This mission was carried out at night, a time we knew the enemy was unlikely to be on the move. It went smoothly. A couple days later, the battle damage assessment came in. It confirmed a body count. This assessment came from an independent source which was as good as having eyes on the ground.

Once I was sure we could work with the guys in gray and that we could give them what they needed, I took a proposal to the operations officer. I told him what I'd found and how I felt we could use it to focus our efforts and start getting some results. I proposed that we set up courier, mail and telephone communications links with my new source of intel and bypass both Group and II Field Force.

The intel would be fresh, I told him. And we could get some good results for the firing batteries. I told him all we needed was a good intel map of the III Corps area, and the guys in gray would provide that for us. We could simply update it as the new intel came in.

By using the assistant operations officer, one of whom was on duty all the time, we could monitor the communications and target the intel as quickly as possible. By using minimum standards for shooting the targets, the officer in charge could generate new and fresh targets from the intel as it came in. On the occasions where the guys in gray had something really hot, they could request the fire mission directly.

We set up the system. It worked beautifully. Within a week, we were placing effective fire on good intel targets. We began to get reports through other channels that confirmed that for once, the S-2 at the battalion level was functioning as it should.

I made it a practice to compare the intel reports from Group and II Field Force with our new map. One morning, we got a report from an infantry unit that had discovered several fresh graves in the middle of an enemy rest area. The report confirmed the bodies showed evidence that the cause of death was artillery. The location where the graves were discovered was about twenty kilometers east of Long Bien. The location had evidence that it had been used as a base camp for a long time. The report indicated the infantry unit stumbled onto it. What struck me the location was the same grid location that we'd fired an artillery mission on two nights before. The target had been generated by our new map.

The report confirmed to me that the raw intelligence was there, but, as it was filtered down through the chain of command it sometimes got lost.

The system worked well, but it wasn't perfect. By cutting out the higher staffers, we were acting on intel before they even knew it existed. A few times, we found out plans were being made to react to some piece of intel after we'd already taken care of it. With the mission involving the base camp, for example, I screwed up an entire brigade-level operation by firing on the objective before the infantry could get to it. Had I not fired on the target, the infantry unit might have captured an entire enemy unit. What they found was twenty fresh graves and an abandoned base camp.

But my higher-ups never knew that I'd cut them out of the action and beat them to the target. At least I don't think they knew. It was 1970, after all, and everyone was used to failure. Maybe they knew and they just didn't care.

I sure didn't care.

That was the fun side of the war, when my chess game helped me make sense out of all that had happened before. I was a combat officer charged with the responsibility of meeting the enemy and defeating him on the field of honor. I liked that role. I felt mesmerized by the power I wielded over my enemies, both those that I hoped to kill and those that I hoped to humiliate. I started feeling like I could stick it in their ass anytime I felt like it. All I had to do was study my map and pay attention to the guys in gray. And if I was lucky, I'd also get a shot at my most worthy opponent.

And nobody really gave a damn.

Binh Hoa
Late spring 1970

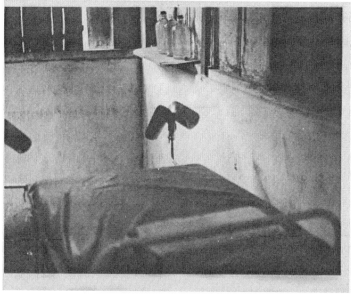

One medical civic action project that did not involve Doc Carr or his staff was the rebuilding of a midwife facility in the Bui Tieng hamlet of Duc Tu District.

There was a tremendous shortage of civilian doctors in Vietnam. There were not enough to staff the hospitals, let alone clinics or other medical facilities. The solution was to train nurses to manage the less critical activities, such as childbirth. That way, one doctor would supervise several midwife facilities in a kind of neighborhood maternity ward setup.

The Bui Tieng facility was designed so the expectant mother could move into the waiting room and stay for several days if necessary. She could stay there before the baby came and as long as necessary afterward. Sometimes the expectant mother would bring her other children. There were facilities for her to cook and sleep. One family would sleep on one bed.

The purpose of our project was to update the facility. My first visit to the ward really surprised me. My hooch was built better than the facility. There were no doors, screens on the windows, and no running water in the facility. The only electrical devices were two single bulb lights hanging from the ceiling, one in the sleeping area and one in the delivery room. The exterior walls were made of one-by-eight wooden boards. The cracks between the boards had never been covered. The facility was open to anything that walked, crawled, flew or slithered. While the facility was clean, it was not clean in the way I thought a medical facility should be clean and sanitary.

The structure was about thirty feet long and thirty feet wide. It had three major areas. The front portion was where the nurse kept records, medicine, and other administrative necessities along with several beds where the mothers stayed. The back portion was divided into two rooms about eight feet wide and fifteen feet long.

One of the rooms was the cooking area and the other was the birthing area. There were no doors; the rooms just opened up into each other.

There was no stove for cooking. They used an open fire. The smoke was vented outside through cracks in the outside wall. The whole place smelled of smoke, even the delivery room. The openings in the exterior walls were designed for windows but there were no windows. They had been covered with damaged wire screens that didn't keep the insects out. The wooden walls were never finished on the inside, so the only covering was the exterior siding. It was a kind of bat-and-batten siding with one-by-eight boards running vertically. The cracks were supposed to be covered with another board, but this second board had never been installed. You could see through the walls. There was nothing to stop the flies from coming in.

An old man lived in a lean-to attached to the exterior of the structure near the delivery room. There was an opening in the wall where a window should have been. This opening was between the delivery room and the lean-to. Every time the old man cooked, the smoke went into the delivery room. He also could watch the deliveries.

There was no running water in the structure; the nurse had to carry what water she could get from a public water tap about a block away. When I first saw the delivery room, there a couple of small jugs of water available. The only other equipment in the delivery room was the bench with the leg stirrups.

After a short inspection of the facility, I went back to the delivery room and stood there looking at the delivery table. I envisioned my own wife lying on the table with her legs in the stirrups, that old man staring at her pussy, his eyes wide, his mouth agape.

I could see myself asking, "What's the matter, old man? You look like you've never seen one of those before."

And he would respond, "I've never seen one like that before; it's a round eye."

I would have to respond, "I wouldn't know, old man, I am afraid of it. I don't look at it, I just fuck it. You of all people should know better than to stare at those. The power of the pussy can make you blind. For Christ's sake, old man, see what that particular one did to me."

I caught myself smiling and then became aware that everyone was watching me. It was not a good time to be smiling. It probably made them wonder what kind of a person I was.

So, no running water and just two light bulbs, no refrigeration for critical medical supplies, and no way to keep the delivery room sterile or private. I was amazed that it worked. But it did.

Nurse Quan was young, pretty and unmarried. She was tied to the facility with a two-year commitment. Despite having no vacation time and only a few hours off each week, she was dedicated. She was expected to take care of all who came in, and that's what she did. She never saw the doctor who was supposed to be supervising her.

I came to look at Quan like I looked at SSG Son. They were special in my mind.

After several meetings, Son and I worked out a remodeling scheme where we would supervise the rebuilding of the facility, using as much local labor as we could. I would buy the materials using battalion funds, and the mayor of Bui Tieng would line up the village's carpenters, plumbers, electricians, and painters.

Best of all, I got to hold the new babies.

After I'd begun spending a lot of time at the maternity ward, I developed a friendship with Nurse Quan. She knew I liked being there and that I liked being around the babies. She found out that I had a little girl and made me talk about her. SSG Son translated for our semi-private conversations. We spent a lot of time talking about children. I did not hide my feeling that my daughter was very important to me, that my daughter was a key to my heart. She understood the brand new babies were a comfort to me.

She wanted to know about American women. She wanted to know about my wife. She asked me if I had a picture of her. I did not. When I received my wife's letter telling me not to write to her anymore, I took her picture out of my wallet. My responses to

Quan's questions about my wife did not contain the warmth for her as they did for my daughter.

My wife was finishing college with my money—money that dripped with the blood from my hands. I resented that she was home safe and living the good life while I lived the highs and lows of war. It made me mad that I was at war because of her and she didn't have the interest to read my letters about my war.

When she asked me questions about American women in general, my responses were a little cryptic. My reference for my answers to her were Kent State, Jane Fonda and hippies. I think she began to sense that I somehow held bad feelings toward them.

When this subject came up, she would listen to Son translate my responses, then she would look at me as if to say, "You are wrong, John. Women are women all over the world. You are not in this alone. Your heart is in the right place. We can help you, if you will let us."

I couldn't even speak to Nurse Quan without an interpreter, but she knew more about me than my wife would ever know.

As the renovation progressed, I realized I didn't have enough money in the budget to buy all the materials we needed to complete the project correctly. So I started looking around for other options. One evening, during my briefing for LTC Mullens, I explained my problem of lack of funds to complete the renovation of the maternity ward. After the briefing I asked the S-4 if he had anything he could give me. He turned me down flat. He didn't even consider the request.

His world involved building and maintaining fire support bases He was responsible for the beans and the bullets and the sand bags and the toilet paper necessary to fight this war. He didn't have the time or the interest in anything not directly related to the battalion. That included maternity wards. His war was far different than mine. It was his responsibility to build fire support bases to protect American soldiers from the enemy. Building maternity wards was far down his list of priorities. LTC Mullens was sympathetic when I brought it up at a staff meeting, but he offered no solutions. I was up a creek.

But after the meeting, the service battery executive officer came up to me and said he could get me any materials that I needed, so as long as I could get someone to load it.

This was in direct conflict with what the S-4—this lieutenant's boss—had told me. But who was I to look a gift horse in the mouth?

My intel sergeant, Master Sargent Wilson, and I put together a crew and solved the problem that very night. Sometimes you get help from the most unexpected sources.

A few days later, the S-4 (Service Battery Commander) called me to his office. He jumped all over me. He asked me if I had stolen the supplies and I told him no. He knew that I had taken them. If he had used the word 'taken,' I would have said yes.

It just so happened that I had taken was the supplies he was going to use to build a massage parlor. The massage parlor was to be a legal whorehouse on our compound. The whorehouse was to be on the compound so that Doc Carr could get control of the VD rate.

Everything was on schedule until I took the supplies. Now everyone was pissed at me because I had fucked up their fucking. The S-4 knew better than to call me a liar to my face; he did not outrank me. I didn't think he knew how I got the supplies, so I decided not to involve the lieutenant. I just stonewalled him instead.

The next day, LTC Mullens called me in and said he wanted to visit the project. The tone of his voice told me he had been talking to the S-4. I took him to the project, let him meet the people and see the babies, the flies, and the old man out back.

I never heard another word about it.

My maternity ward project was completed in late August, about a couple of weeks before I was to rotate back to the States. When we walked away, the ward was fly-proof, and the delivery room had running water, electricity and doors. It was not sterile, but it was clean. And the old man in the back had a better place to live.

There was a decent kitchen and the recovery room was livable. Nurse Quan had a refrigerator for the drugs, and Doc Carr had spent time with her to make sure she was up to date with the procedures. We had done a good job. We had worked with the Vietnamese and we had built something that was useful, and in doing so, we all were better for it.

I was very proud of my maternity ward and I was impressed with those who had worked with me. In the end, I was the one who walked away with a lesson in life. I took away so much more than I'd given.

On my last visit to the ward, Nurse Quan gave me a small tea set she had made out of eggshells and paper. It was a present for my daughter.

As she handed it to me her eyes said to me: Drink from this, my small but heartfelt gift, and you will thirst for life.

Battalion Compound
June 1970

I had a pet. Well, maybe not a pet—he was more like an experiment. I named him John Fitzgerald Kennedy. I can't remember why. I kept him in a water glass on a ledge in my hooch. Under the glass I placed a sign, "Ask not what your country can do for you, but what you can do for your country." I don't know why I did that, either.

John F., as I called him for short, was the biggest cockroach I ever saw. His mistake was getting caught in my hooch. By the time he came along, I had found cockroaches in my food, in my clothes, in my bed and in the shitter; I found the sons-a-bitches every goddamned place I tried to hide. They were pissing me off.

It so happened that when I found John F., I had a glass of Coca-Cola in my hand. I had heard that if you placed meat in Coke, the Coke dissolved the meat. That's how I came up with the idea for this experiment. When I discussed it with John F., he was reluctant. But I outranked him.

I promised John F. that if he survived the experiment, we could be good buddies. I coaxed him to latch onto the end of a pencil, then placed both him and the pencil in the glass of Coke. After watching him a while to make sure he didn't pull a fast one on me and climb out, I started the experiment. It was an open-ended type of experiment, meaning that I would conclude it when I felt I had enough data to make a rational decision about the power of Coke to eat flesh.

Each day I would check John F's condition. Each day, as he clung to the pencil with about half of his body submerged in the Coke, he would look up at me as if to say, "You asshole." But he seemed to be doing all right. After three days I decided I should take him out of the glass and see what effects the Coke had on him.

He was just fine. He started to crawl off, as if saying to me, "See you later."

"You fucking cockroach."

I stepped on John F. Kennedy. His body crunched and splattered beneath my VC sandals.

Nothing was sacred anymore.

South Vietnam
Summer 1970

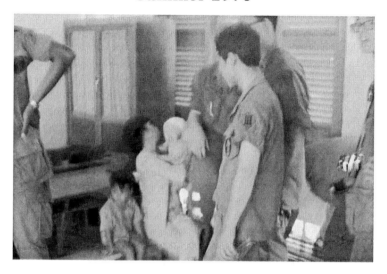

During his campaign, President Nixon promised that he would reduce America's involvement in the Vietnam War. His plan was to transfer the responsibility for fighting the war to the South Vietnamese while America took on a support role. The name he gave to this change was Vietnamization.

Vietnamization had several meanings to me during my tour of duty. There was a crash program for us to train our sister ARVN battalion about the 155-towed artillery howitzer. It was our job to teach these South Vietnamese soldiers how to act like an American artillery unit in combat. We did not know when our battalion would be deactivated, but we knew it would be within the next year or so. Our ARVN brothers had to be ready to take on the fight when that happened.

The aspect of Vietnamization that made it worthwhile for me was the part that happened after the bullets and the artillery rounds had been put away, when Americans and Vietnamese got together and tried to build a new and exciting country. For hundreds of years Vietnam had been governed by foreigners. First it was the Chinese, then it was the French, then the Japanese, and then the French again. After the French, the Americans came along.

It was the Americans who wanted the South Vietnamese to govern themselves. But self-government was not something that came about overnight. Bombs and bullets do not win wars. Wars are won when minds are changed. It was the Americans who introduced the concept of civil rights to the common person in Vietnam. After hundreds of years of servitude, the mindset of the people of South Vietnam was that of servitude. There was no concept of self or individual in the Vietnamese language when America showed up. That mindset had to be changed through education and by building a sense of ownership. Building that sense of ownership was something that I wanted to help with, and with our civic action projects, we had the tools to do just that.

Our battalion's civic action initiatives focused mainly on providing medical care. We had been blessed with a Black Panther battalion doctor, Doc Carr, and a group of medics who were really motivated by this work. Our MedCaps (Medical Civic Action Program) facilitated our relationship with our sister ARVN battalion by giving them an opportunity to experience the medical profession outside the restrictions of military compounds. It was the MedCaps that provided our medical section the opportunity to train our ARVN brothers in the process of medicine. And in so doing we helped people; we helped villages and hamlets to deal with injury and disease.

This was cool stuff. This is what I volunteered for.

It was my responsibility to organize and supervise these projects. Because of the battalion's long record of involvement in civic action projects, the ARVN had assigned SSG Son to us. And because of my responsibilities, he was assigned to work with me.

It was during these MedCaps that Doc Carr trained the ARVN Battalion medical staff. By carefully selecting the location of the MedCaps, we and our South Vietnamese counterparts would try to win the hearts and minds of the people in the village while accomplishing our major mission.

Some villages were too dangerous, some villages were clearly in support of the South Vietnamese authority, and most villages were somewhere in between. It was the in-between villages that needed

MedCaps. Medical personnel from both battalions would go to the selected village and set up a clinic that saw patients two days a week. They treated any medical issue the villagers showed up with, from gunshot wounds to tuberculosis. It was during these MedCaps that Vietnamization took on a civil aspect. The goal was that just maybe we could win some friends for our experiment with democracy in South Vietnam.

For all the terror inherent in simply being Vietnamese in 1970, those villagers' ability to retain their social order showed a resiliency that amazed us. We knew the village was under tremendous pressure from the South Vietnamese government to conform and support them. And that there was just as much pressure from the VC for them to rebel. Sometimes the threats, the intimidation, the propaganda and the acts of war of both sides were there at the same time. Sometimes as one would subside, the other would intensify, but always the pressures were daily. Our MedCaps were peaceful acts on the part of the government that intended to blunt the VC propaganda.

There was always a risk that the VC would win.

I remember vividly the mistrust of the villagers as we first tried to convince them we came to help them. We were also the ones tearing their world apart. It was us who could bring death and destruction down on them at a moment's notice. They blamed us for doing just that. Sometimes the tension was so thick you felt smothered by it.

I first noticed the mistrust in a *mama san* standing in line waiting to see Doc Carr. She was holding her baby son. His head was covered with black tar and he had open, oozing holes in his scalp. When she looked at me, I saw hate and desperation in her eyes. I asked SSG Son why she wanted to see Doc Carr. He conversed with her and turned to me.

"She feels we are her last hope for her son." SSG Son explained that she had been taking him to the local Chinese doctor to fix her son's skin problems. But the problems were getting worse, her son was getting sicker and sicker. She is afraid she would lose him. She had no other place to turn.

The Chinese doctor had treated the boy by putting tar in the open sores. The skin infection kept growing and the boy's tolerance for the infection had broken his health down.

Doc. Carr examined the boy, asking his mom some questions through SSG Son. Then he and two medics, one American and one Vietnamese, started working on the boy. First they had to remove the black tar from the open wounds. Sometimes they had to cut the tar away. Doc talked to Son and he translated to her so that she knew what they were doing and why they were doing it.

It was brutal to watch. Mom was crying, the boy was crying and struggling. The medics were working hard to hold the boy and provide the medical equipment and supplies Doc needed. The smell made my stomach turn over. I had to stop watching before I lost it.

Once they got the wounds free, they cleaned and sterilized them. Then they packed the wounds with gauze and stitched them closed with a little gauze protruding to act as a wick to let them drain. They gave the boy a shot of something to fight the infection.

Once they were finished, Doc. Carr sat down with the boy's mother and, through SSG Son, explained to her that the cause of her son's problems was contaminated water. He took his time to show her how to administer the pills the boy needed to take every day. He gave her two quarts of sterile water and told her not to use anything but the sterile water to clean her son's head. Under no circumstance should she try to clean the wounds.

Then he told her, not asked her, to bring her son back to the clinic the next time it was open. (In two or three days).

I don't know if Doc Carr noticed the mistrust in her face. He probably didn't, because he was a busy guy. I did notice that her face seemed a little softer after he finished talking to her.

A couple of days later, she and her son were back. Doc Carr examined the boy and found that the wounds were healing as expected. The medics extracted the gauze from the wounds and finished stitching them closed. Through SSG Son, he complimented her for taking good care of her son. He took time to show her how to keep the wounds clean and sterile. He gave her some more pills, sterile water, creams and bandages, and explained how to use everything. He told her why it was important for her son to get the pills as directed.

I could see in her face that Doc Carr was becoming her hero.

The next week she was back, and the medics removed the stitches from her son's scalp. There would be some fine, small scars where the sores were, but nothing more. Her son was active and full of life. Doc Carr was still a busy guy, but he had time to hold this young boy and receive the thanks of his mom, with smiles all around.

That look I saw on her face that first day was gone.

—

Even though I understood why they were suspicious of our intentions, it wasn't easy to show up day after day to help people who wouldn't even smile at you.

Now and then, a particularly successful medical treatment could result in honest appreciation from the patient and sometimes even the whole village. It was like a small light would flicker to life in these moments. They never lasted long, and we learned to savor them.

The MedCaps became my real reason for being a soldier. The bullets were still a major part of my job, but what I really craved was to be the loving, compassionate being who was willing to give his all so that a few could live. Typically, in one day, our MedCap would treat two or three hundred patients. We treated everything from gunshot wounds to chronic diarrhea. We treated them before they were born, when they were born, and when they had grown old and feeble.

It was my little way of liberating Dachau, of giving the chocolate bar to the child, of saving the old folks from the ravages of war. It was my way of being John Wayne.

Twice a week, we would load up a couple of three-quarter ton trucks with medics and supplies and go to a small village close to our sister battalion's headquarters. We'd set up in a school building and treat any and all comers, no questions asked. On these projects, my only real function was to make sure Doc Carr and his medics had transportation to and from the project. Once we got there, I just sat around watching the Doc, the medics, and Son do their work.

The Americans and the Vietnamese worked well together because we shared the same goals. A Wyoming cowboy, a Black Panther doctor, a North Vietnamese interpreter, and South Vietnamese soldiers, nationals and revolutionaries got it done. With just a little compassion, we discovered we could all make it work. If only we could have translated that cooperation to rest of the war.

We did have our problems. It wasn't long before we found ourselves in direct competition with the local medical establishment. We understood we'd have to prove ourselves at first. No responsible parent would subject their health or their family to treatment by an interloper without good reason to believe no harm would be done.

We were challenging an established system of medicine, in the form of a Chinese doctor who carried on a family practice established generations before. His methods were familiar to them; ours weren't. So at first, we only treated minor illnesses and injuries. You could see the skepticism etched onto their faces in photos we took for documentation.

And it was not a one-way street. We didn't carry arms during our MedCaps, so we knew that we were easy targets. There was always the possibility that the V.C. could put a stop to our mission any time they chose to do so. Any of our patients could have caused us harm. One grenade could have destroyed us all. There was no way to tell which villager might be there to cause us harm. But there was no good that could come from dwelling on this risk, so we didn't spend a lot of time or effort thinking about it.

In a way, we had to operate on blind faith that the people we wanted to serve would not hurt us. On the other hand, they didn't know what to expect of us. In the beginning of a MedCap mission it was up to us to cross a kind of a no man's land between us and them before we could do our job.

There were many stories about unlucky MedCaps teams who had tangled with the NVA or VC. These stories were never far from my mind since I was charged with protecting the doctor and his staff. I never knew how to do that, other than to rely on blind luck.

Twice a week for several weeks, our little convoy would roll into the village, stop at the school, and set up operations. It did not take long before we found hundreds of people waiting on us at the school. They were there to receive medical treatment from the American doctor and his people.

As we became a regular fixture around the village, the people began to accept Doc Carr and his team, and we started seeing patients with more serious conditions. One day an emaciated old man with tuberculosis came for treatment. The American Army did not have a problem with tuberculosis in 1970, but somehow, Doc Carr found what he needed to treat the old man. I never knew where he got the medicine, but he managed to find enough of whatever he needed to treat the old man for several weeks. Everyone saw the old man begin to respond to the treatment. As he improved, the environment in the village warmed for us.

Then one day, a desperate mother brought in a baby suffering from chronic diarrhea. The Chinese doctor had been treating the infant with no success. The baby was close to death by the time she got him to us; he was nothing but skin and bones. With Son interpreting, Doc Carr told the mother that her baby's condition was too serious to be treated at the clinic, and that he'd have to be transported to the battalion's aid station to be fed intravenously and monitored around the clock. He told her that he was afraid he could not save the baby's life otherwise, and that she would be allowed to stay with her baby. She was ready to try anything, even trust an American. Doc Carr loaded her and her baby into one of our three-quarter-ton trucks, transported them to the aid station, and began treating the baby.

Things were going well until LTC Mullens found out we were treating a Vietnamese baby in the aid station. Neither I nor Doc had thought to request permission from Mullens to bring the baby onto the battalion compound for serious medical treatment. Mullens went ballistic. He called me into his office and dressed me down. He informed me that if the baby died while in our care, it would cause an international incident.

That was something I had not thought of.

LTC Mullens ordered me to get the baby and his mom out of the aid station as quickly as possible. I went to the aid station and told Doc what Mullens had ordered me to do. Doc said the baby would die if we withdrew the treatment, and that our only hope was to find a civilian hospital that would agree to assume the treatment.

Doc had a contact in a Catholic Hospital in Saigon. He made the call to his contact, and Son and I traveled to the hospital to make the arrangements to move the baby and his mother. On the way to the hospital in Saigon, I explained what was going on and why to SSG Son. It was Son's opinion that if the baby died, there would be no incident. Everyone knew we were trying to save his life. But Son understood that we had no choice.

With as many trips as I took into the real Vietnam, it was inevitable that I would be exposed to its ugly side.

The Vietnamese doctor at the hospital showed us into the children's ward so that we could see the operation. We walked by a room with a bunch of young children and several babies inside. I was struck by the fact that there was no noise. Most of the children looked lethargic. Several were lying on the concrete floor, not moving. As I watched them for a while, I noticed that every child was of mixed blood. One was sitting in a far corner and banging his head against the wall. There were no adults in the room. None of the children played or spoke. They just stared, back at me or at nothing.

They were dying.

I asked Son to find out what was going on in that room, why the children were segregated like that.

"Vietnamese don't like mixed kids," he said. "It is very bad for a Vietnamese mother to have a mixed kid. She will be kicked out. Vietnamese want to keep their blood pure."

"What do they do, bring the mixed kids here?"

"Yeah. Mixed kids don't stand a chance in most villages."

"It looks like they don't stand much of a chance here, either."

Several of the mixed-blood kids were black. I asked Son about them.

"The black ones are worse off than the white ones," he said.

I couldn't get my eyes off those kids. They had given up and laid down to die. The lack of love was too much for them to cope with. Life was finished for them; the spark was gone you could see it had left their eyes. I had seen dead eyes living, now I was seeing living eyes dead.

I could not help thinking that some of these kids were from the same guys demonstrating against racism in America. They come over here, make a baby, and then abandon it to a system that is more racist than anything they could imagine.

I could understand the riots, the Black Panthers, and Dr. King's dream. The lack of leadership both at home and here in Vietnam helped explain the drugs and the racial tensions in the Army. But I could not understand how they could just leave their own flesh and blood to rot away like this.

—

Doc and the medics moved the baby and his mom to the Catholic hospital using the battalion ambulance. When I got back, I reported to LTC Mullens that we had moved the baby to a Catholic hospital in Saigon. He listened to my report but did not respond with a question or a comment. He just thanked me for the report and dismissed me. As I left his office, I was not sure how I felt.

Later, after the nightly staff meeting, I asked Doc if he thought he could have saved the baby. He said he knew he could have. The baby had already been showing good signs when they put him in the ambulance. Doc said he had no idea what happened to the baby once they transferred him to the hospital.

A couple of weeks later during a clinic, I asked SSG Son to find out what happened to the baby.

It took him several minutes to find someone in the village who had information about the baby. He reported to me that the baby had died in the hospital.

I told Son not to tell Doc or any of the medics what we had learned. I felt that it was better that they did not know. LTC Mullens never asked about the baby, and I never brought up the subject to him.

You have shown them your way is a better way than theirs, my friend. What if you can't give your way to them? You are not going to be here much longer. What if you don't come through? What will they do then? Is this all just a big lie?

What would happen when we ended our mission in Vietnam? Would they become skeptical of their traditional medicine only to find our medicine was no longer available to them? Were we making a mistake? Had we created an impossible condition for them?

Maybe we had. But it was good to see them smile.

Battalion Compound
Summer 1970

As rewarding as my work with civic action programs and the guys in gray had become, there was another side of the war that took on an ever-increasing blackness. Our compound was relatively secure from attack, meaning we did not take a lot of incoming. Most of the incoming we did get was ineffective. As the battalion security officer, I wasn't concerned with what went on outside the wire. It was what went on inside the wire that caused me to sleep with a loaded .45.

It was interesting how this battalion could function so well in the field, day or night. However, at night when the battalion compound was supposed to be secure, it became a place for a different animal. The booze and the dope transformed the troops into a new social order that could tolerate little interaction with traditional military values. Those who abstained could ignore this new order only as long as they could avoid confronting it. If I hadn't been the battalion security officer, I would have been one of them, slinking around with my head down, trying to be invisible.

We had a saying that if the monkey was out, it would soon be on your back. What that meant was that there were a number of roadside stands around Long Bien and Bien Hoa. These Vietnamese vendors sold all kinds of wares, from ice cold Coke to black market stereos. If the stand was displaying a monkey, ostensibly the proprietor's pet, it meant there were drugs for sale.

The monkey was out, and it had found a comfortable perch on our battalion's back.

That monkey was never more evident than when we received incoming. Incoming usually generated a light show of red and green tracer bullets. Some of the tracers coming in and some going out. I figured it was one of my duties as battalion security officer to check the perimeter defenses. The OD (officer of the day) exercised command responsibility for the security force. I couldn't interfere with his authority, but I could observe the action. I figured he was a busy guy and could use the help. I would identify the most critical area, then go to the duty posts to assess damage and to check the troops. When things settled down I'd work around the entire perimeter, checking all the bunkers, towers, and command posts.

Sometimes I'd find troops sitting on the top of the bunkers enjoying the light show. If they didn't respond to my voice or if their response to me was that of an idiot, I'd pull them down, get them inside the bunker, and check to see if they even knew where they were. If I found someone in the bunker who could stand, speak anything but Vietnamese, and show me where the wire was, I figured I was lucky and I left them alone.

I divided the security troops into two potential groups: those who were clean while on duty and those who were not. If I found a troop asleep while on duty—a court-martial offense—and if I was able to ascertain he was clean, I would take his post until he woke up. Nothing more would be said.

But if the security troop was doped up, he could not be counted on to do his duty. I would contact the NCOIC (non-commissioned officer in charge) of the security force to relieve him from duty and replace him with a new troop. I didn't care what they did on their own time, but I did care when their stupidity threatened my own security.

The extra duty of command of the security force was shared by all the officers of the battalion staff. This duty was known as officer of the day, or OD. It worked out that I was OD about once or twice a month. That meant I was in charge of the security force for the compound during the night hours. If any problems, combat or otherwise, came up, the OD was to handle them until the regular command structure took over again.

My first conflict inside the wire occurred while I was acting as OD. The OD duty started with guard mount at 1700 on the evening before the security force would begin their detail. During guard mount, I'd give the guards their orders for the day, a short intelligence briefing, check their weapons, and test their knowledge of their duty assignment. It was at guard mount that the OD was to determine whether the guards were present and capable of handling their assignments.

As I approached the parade ground where we held guard mount, I looked for the NCOIC. I had the roster and I knew that my NCOIC was a staff sergeant. His name was SSG Barnes, and he was new to the battalion. I didn't see him, but I spotted SGT Brown and asked him where the NCOIC was. Brown said he'd never met SSG Barnes and had no idea where to find him. I ordered Brown to form up the guard detail and prepare for guard mount, expecting that Barnes would be along shortly. I would cut him some slack because he was new.

Guard mount took about thirty minutes. When I was done, Barnes still hadn't shown up. I instructed SGT Brown to find SSG Barnes and relay my order for him to report to me at the TOC. I had a few more intel targets to plot and coordinate with the firing batteries, so I walked to the TOC and thought nothing more about the situation.

About two hours later, after finishing the targeting, I went into the communications center for the security force. Brown was there testing the communication networks for the bunkers. I asked him if he had found SSG Barnes.

"Yes, Sir."

"Did you tell him to report to me?"

"Yes, Sir."

"So where is he?" I asked. Brown didn't know. *Well, hell*, I thought. *This is a new one.* "Okay," I said to Brown, "This time you go find Barnes and tell him to report to me on the double. Tell him if he's not here by 2000, I will consider him AWOL."

"Yes, Sir."

About 1945, SGT Brown walked into the communications center.

I looked up from the book I was reading. "Did you find him?"

"Yes, sir. He's in the NCO club. Drinking, sir."

"You're shitting me. Did you tell him to report by 2000?"

"Yes sir."

"Did he understand you?"

"I think so, sir."

This pissed me off.

I waited until 2000. When Brown didn't show, I put on my helmet and flak jacket, then checked the ammo in my .45 caliber pistol. I ordered Brown to give me a bandoleer of M-16 ammo for my rifle, just for effect. Everyone in the commo center was watching me. I made a statement for everyone:

"I'm going to the NCO club. I will be back shortly."

As I went through the door of the communications center, I slapped a magazine into the M-16.

Just for effect.

When I entered the NCO Club, the first person I saw was the battalion sergeant major. He asked me if I wanted a drink.

"Sorry, Sergeant Major, I'm on duty. Is SSG Barnes in here?"

"Yes, Sir." He pointed to a table in the middle of the club, next to the bar. "Right over there, Sir."

"Thank you, Sergeant Major." I turned toward the table he pointed to, took a moment to take in my surroundings, and began walking. As I approached the table, I noticed my intelligence sergeant setting with Barnes. They were both drinking hard liquor.

I pulled back the receiver on the M-16 and let it slam forward, loading a round into the chamber. The noise sounded like an explosion inside the club. Everyone stopped talking and looked at me. Once I was sure everybody knew that I was there, I took my thumb and switched the M-16 from safety to semiautomatic.

I lowered the muzzle of the M-16 until it pointed directly at Barnes. Right between his eyes, so close I felt the muzzle bump against his forehead. His eyes were as big as saucers.

"Sergeant, I am CPT Martin, OD. I am here to inform you that you are AWOL from your post in time of combat. You have disobeyed direct orders while in combat. You have abandoned your post while in combat, and now you are drunk while on duty. I am of a mind to blow your head off right here. Do you have anything to say for yourself?"

"Sir," he slurred; "I didn't know I was on duty tonight."

"Bullshit. You've been told twice that you're on duty. You have five minutes to report to the TOC or you are a dead puke. Do you read me, Sergeant?"

"Yes, sir."

He didn't make an effort to move.

I yelled at him: "Get your ass out of here, now!"

"Yes, sir."

As he got up, I could tell he was staggering drunk. He'd be no good tonight. As he stumbled toward the door of the club, I laid the M-16 on the table with the muzzle toward my intel sergeant.

"Master Sergeant Wilson, what the fuck are you trying to pull? You know better than this."

He didn't answer. I just stared at him for a few seconds. "We'll discuss this tomorrow. "

Then I picked up the M-16 and walked out. I couldn't help smiling as I left the club, I was actually having fun.

Those good old boys might be wondering what made me tick about now.

Before returning to TOC, I checked a few of the bunkers on that side of the compound. It was pitch dark when I got back to the TOC. I could see someone walking just ahead of me as I entered the communications room, but I couldn't tell who it was. When I got closer, I recognized Barnes. Everyone in the room was looking at us as we entered.

Barnes must not have known I was behind him. "Howdy," he said. "Does anyone have a joint I could bum?"

I couldn't believe what I had just heard. The men stared at me over Barnes' shoulder. I was as shocked as they were. If the situation hadn't been so serious, it would have been funny.

I came unglued. I grabbed him with my left hand and spun him around. Then I placed my M-16 sideways on his chest and shoved as hard as I could. He lost his balance, went sprawling, and slid headfirst into the bunker wall. I ran over and started to hit him with the butt end of my M-16. SGT Brown jumped between us and grabbed me. A couple of off-duty guards picked Barnes up while Brown calmed me down.

We made eye contact and he said, "Sir, please don't do this."

It took a few seconds for me to recognize what Brown was trying to do for me, and when it finally sunk in, I shook my head to let him know I understood.

I stepped around Brown and looked down at Barnes. "Get your ass up to the flash base tower, now," I yelled down at him. "I want you to complete a communication check every fifteen minutes for the rest of the night." As an afterthought, I added, "If you miss one goddamned check, I'm going to come up there and throw you out of the tower. Do you understand me, Sergeant?"

"Yes, sir." He wobbled to his feet. "Where is the flash base tower?"

"You son of a bitch, I ought to kill you right now!"

I realized that I meant it. I *could* have killed him. That cooled me off some. I looked over at Brown. "Come with me, Sergeant," I said. "I need someone to keep me from killing this son of a bitch."

We took Barnes to the flash base tower. It was a platform sixty feet above the ground from which we watched for incoming mortar rounds so we could direct effective counter mortar or suppression fire at the enemy. If we could see the flash as the round left the mortar tube, we could establish a direction to the tube. If two flash base towers were able to establish a direction to the mortar tube, we could triangulate on the tube location and attack the enemy using the mortar tube.

Since it was our enemy's number-two target, right behind the number-one target, the TOC (Tactical Operations Center), the flash base tower was not a nice place to be even when the sun was out. At night it was spooky.

The tower swayed as Barnes made his way up the narrow, slick ladder toward the platform.

SGT Brown said, "Sir, you're still trying to kill him, aren't you?"

"Yes, Sergeant," I said. "I think I am."

After Barnes finally made it to the platform, Brown and I turned to go back to the TOC. We talked about Barnes. SGT Brown believed that if he'd done what Barnes had done, he'd be court martialed. Because Barnes was a Staff Sergeant, Brown didn't think Barnes would get punished.

I had to agree with him. The good old boys would take care of Barnes. I figured it was up to me to do something. This situation should not stand unresolved.

When we got back to the communications room, I instructed Brown that I was going to finish checking the perimeter. I took Brown aside so we could talk privately.

"When I get to Bunker Eight, I will call you and instruct you to call Barnes. Tell him to report to me in five minutes in full combat gear at Bunker Eight."

"Sir, SSG Barnes doesn't have his combat gear. There is no way he can report to you at Bunker Eight in five minutes."

"I know that, Sergeant."

"Gotcha, sir."

When I got to Bunker Eight, across the compound from the flash base tower, I called Brown and told him to put my plan into action. I instructed a nearby guard in Bunker Eight to wait outside and report to me when SSG Barnes arrived.

Before the guard left, I asked him what time it was. It was 21:35. I had a watch, of course, but now I'd have a witness for what was about to happen.

While I waited for Barnes, I went through the motions of inspecting the bunker and testing the guards on information such as their password, their fields of fire, and the location of their additional ammo.

I chose Bunker Eight because it was one of three bunkers at that location. The other bunkers were within earshot, so I could be reasonably sure all the guards in all three bunkers would hear what would transpire between Barnes and me. It was my intent that the grapevine would alert the entire battalion as to what happened by noon the next day.

The guard I'd sent to wait for Barnes stepped into the bunker and told me Barnes had arrived.

"And what time do you have now, Specialist?" I said.

"I have 21:53."

"Thank you," I said.

As I exited the bunker, I spotted Barnes on one knee, head hanging, and helmet on the ground. He had run most of the way across the battalion complex and it was clear he was hurting, but he was in full combat gear. At least he'd gotten something right.

"Stand up, Sergeant."

"Yes, sir." He didn't bother trying to hide the contempt in his voice.

"Now, you listen to me, Sergeant. I am going to give you several orders. You better get every one of them correct. Do you understand?"

"Yes, sir."

"By the way, Sergeant, you're late. You disobeyed another order." I could see he didn't seem to care. He was still drunk.

I said. "Sergeant! A-tensh-hut."

"Abooout, face!"

"Poooort, arms!"

"Forward, march!"

"Double time, march."

"Hup, two, three, four. Hup, two, three, four."

"Guide on the road, Sergeant."

"Sergeant, for your information, you're going back to the tower."

"Hup, two, three, four. Hup, two, three, four."

We double-timed about halfway back to the tower before Barnes began to stagger. I leaned closer and yelled in his ear keep him going. But at last, he stumbled and fell face first on the ground. His rifle squirted out from under him as he hit the ground. It ended up about three meters in front of him. He didn't move. I assumed he'd passed out.

Before I knew what was happening, I was standing over him with my .45 in my hand. I locked and loaded a round and pointed the pistol at his head.

After a moment, I lowered the pistol and walked over to the edge of the road and sat down on some sandbags. Barnes still hadn't moved. I sat there for a few moments, waiting for the tension inside me to abate, but it didn't. So I got up, walked over to where Barnes lay, and kicked him as hard as I could in his rib cage. He just grunted.

I felt a lot better. I went back to the sandbags and sat down.

A few minutes later, Barnes began to come around. I ordered him to stand, which he did. But he could not straighten up. He complained that his side hurt. I marched him back to the tower and told him to get back up there. I watched him crawl up the ladder and disappear into the tower.

Then I returned to the TOC. As I entered the commo room, I noticed that the off duty guards were all asleep on cots and SGT Brown was monitoring the telephone switchboard. He turned and asked me what had happened.

I ignored the question. I told him to wake me if something happened. I laid down on an empty cot and fell dead asleep.

The next evening at the battalion staff meeting, Doc Carr briefed us on that week's VD rate and listed our casualties. Then he told us that a new staff sergeant had been injured while on guard duty the night before. LTC Mullens asked how it had happened. The doctor replied that the staff sergeant had slipped while climbing down the ladder at the flash base tower and broken a rib. Mullens suggested that we should make the ladder a little safer.

As the commander and the doctor talked, the battalion sergeant major turned in his chair and smiled at me. I shrugged my shoulders and looked at him as if to say, hey, don't look at me; what the fuck do I know?

There was no further talk about safety improvements to the ladder on the flash base tower.

The Mushroom
Summer 1970

One of my first chores on this particular day was to register Bravo Battery. Its mission was to support the 25th Infantry Division conducting clearing operations in the Mushroom of the Saigon River. This was just north of the Bo Lo Woods and on the northwest edge of the Iron Triangle. It was called the Mushroom because the Saigon River cut off a piece of land that looked like a mushroom when viewed from the air.

The operation was designed to sweep the area north of Cu Chi, west of the Saigon River, and the west side of the Iron Triangle. Since they had moved into the Mushroom, Bravo Battery had been probed almost every night. They'd been working all day and fighting all night, with little or no results. Everyone's nerves were getting short.

The 155 mm howitzer could be a very effective defensive weapon for some targets. The blast was powerful enough to change the momentum of an attack in one split second. The shrapnel effect from one 155 mm artillery round was designed to be lethal.

Targets in defilade, where the enemy is protected from direct observation and can't be effectively attacked with flat trajectory weapons like rifles or machine guns, were a particular problem in defending the fire support base. These weapons could not reach targets located in holes or behind logs, trees or rocks.

To solve this problem, the firing battery developed a tactic they called "Killer Junior." It was a method of direct fire at close range at targets that were protected either by a foxhole or an intervening hill. Killer Junior worked by mating a 155 mm artillery round with a time fuse.

Targets close to the fire support base were usually enemy soldiers probing for a weakness in the defenses or preparing for an all-out attack. By using the fanning effect of the shrapnel from the artillery round, Killer Junior could engage targets you couldn't see.

When artillery rounds explode, they are designed to send the major portion of the shrapnel out in a line ninety degrees from the line of flight. Kind of like an umbrella. This shrapnel effect can dig out bodies in foxholes or ditches or depressions in the ground when these same bodies are protected from direct observation or direct fire.

Because of the size and the power of the 155 mm artillery round, if it goes off too close to the howitzer there will be some shrapnel in the battery area. A minimum distance for safety is approximately two hundred meters.

Killer Junior used a time fuse which would allow the round to travel to the target and go off when it was directly above it. The distance above the defilade position, the enemy position protected from our line-of-sight weapons, would be from two to twenty meters.

Killer Junior was used in defense of an attacking enemy or as harassment fire to keep the enemy off balance. It could be fired at suspected assembly areas at different times during the night, making those areas less attractive for future assemblies. If properly used, Killer Junior was a very effective method of protecting the FSB (Fire Support Base).

Bravo had been using it every night in an effort to deny Charlie the ability to mount a coordinating attack. On one particular night, there had been a probe and the response was to use Killer Junior. It was the job of one of the howitzer crew members to set the time fuse to the proper number of seconds of flight time and install the fuse into the projectile. And, he had to do it in the dark. A towed 155 mm howitzer position cannot be an enclosed position. The howitzer crew is exposed to the weapons of combat.

On this night, he cut the fuse too short. Maybe he was in a hurry, maybe he buckled under the stress, or maybe he was dead tired. The reason is not important. The round exploded too close to the howitzer. The base plate of the projectile came back into the howitzer position and hit another crew member in the head.

When the kid realized what he had done, he cracked. He went over to a bunker, found a .45 pistol and put it to his head. One of the other members of his section saw what was happening and grabbed the kid's arm just as he pulled the trigger. The bullet missed the mark and ricocheted harmlessly off the howitzer.

That night, the section lost two good men, one with a shrapnel wound to his head and the other with psychological damage to his head.

The whole battery was tired. They were frustrated because they could not effectively strike back, and the frustration was making them mean.

—

We had two choices for flight altitude if we didn't want to get shot down. Fly at two thousand feet where the rifle bullets started losing their punch, or fly as close to the ground as possible so that we came and went too fast for a rifle to point and fire. On this summer morning in 1970, we had chosen to fly low, just above the treetops. My Bird Dog was sweeping across the jungle canopy in the bright morning sun during a day-long mission.

As we approached the Mushroom from the south, I spotted a platoon of infantry dug into a position opposite the FSB, across the river to the south. Their position was too obvious for me to be overly concerned. But I decided to have Bravo Battery's Fire Direction Officer see if he could get clearance to fire on the location. I called the fire direction center at Bravo using the Bird Dog's radio and requested the FDO (Fire Direction Officer) to request all affirmative air and ground clearances for the grid location of the platoon. By getting the clearances from all the various agencies, American and Vietnamese, we were assured that there were no friendly troops or civilians in the target area. So we could assume anyone that *was* in the target area was unfriendly.

The grid location I gave to the FDC was the center of the platoon position. I identified the target as being a platoon, dug in. "Three Four" was the call sign of Bravo's Fire Direction Officer, One Two was my call sign.

"Ranger Three Four, this is Ranger One Two, over."

"Ranger One Two, this is Ranger Three Four, over."

"Three Four, this is One Two, see if you can get all affirm air and ground on grid XT580315, over."

"One Two, Three Four, Roger, state the nature of the target, over."

"Three Four, One Two, I have an infantry platoon dug in, open foxholes, over."

"One Two, Three Four, are you calling a fire mission, over."

"Three Four, One Two, not at this time, I want to check this out first, over."

"One Two, Three Four, Wilco, out."

In less than three minutes the FDO called me back.

"Ranger One Two, this is Ranger Three Four, over."

"Three Four, this is One Two, go, over."

"One Two, Three Four, we have all affirmative air and ground. I have two batteries ready to go, over."

"Three Four, One Two, I understand. Something isn't right here, check your clearance again, I want an area cleared one klick on each side of the target grid. I am going to check out the target again, over."

I instructed the Bird Dog pilot to fly over the target again. I was nervous because the FDO was starving for a fire mission. It was an angry, frustrated, and humiliated beast that I had a leash on. Once I took that leash off, whatever was down there on the ground would soon be a memory.

"One Two, Three Four, you are cleared for one klick each side of the target grid, what effect do you want on the target, over."

He was practically begging me.

"Three Four, One Two, no fire mission yet. I want to get closer and check this out. I don't like the feel of it, standby, over."

"One Two, Three Four, over."

"Three Four, One Two, wait, over."

We flew over the target at 2000 feet and nothing happened. I instructed the pilot to get down on the grass and fly over the target knowing full well that if I was wrong, we would get our asses shot off. Down we went. Up went the pucker factor. All the while I was hearing, "One Two this is Three Four, over... One Two, this is Three Four, over... One Two, this is Three Four, over..."

Under my breath, I growled, "Goddammit, I told you to hold on!"

As we flew by, rather than seeing AK-47s pointed skyward, we saw men watching curiously and waving. Most of them were casually carrying M-16's or M-79's. They were a platoon from the 25th Division.

They were Americans.

As we gained a little altitude so that I could see the Fire Support Base and the platoon on the ground below me, I got mad. Not just because the Fire Direction Officer kept nagging me over the radio, but because some asshole had about caused me to kill a bunch of Americans. I was mad and scared all at the same time.

So many lives could have been lost, had it not been for a hunch.

"Three Four, this is One Two, I don't know who the hell you got your clearances from, but you had better check the dumb fuckers again. That target is American, over."

"One Two, Three Four, there are no American units in the area, over."

"Three Four, One Two, bullshit, how do you explain the blond hair and the black skin, then?"

"One Two, Three Four, there are not supposed to be friendlies in the area, over."

"Three Four, One Two, I have been within ten meters of the target and I say they are American. If you don't believe me, you come out here and see for yourself. I will not authorize a fire mission on this target. I say again, no fire mission, out."

"One Two, Three Four, over."

The FDO decided not to shoot the mission. I instructed the pilot to fly back over the platoon, low and slow, so I could drop them a message to warn them that their higher did not know where they were. I put the message in my canteen cover and dropped it as we flew by. Then I tried to forget about the entire incident. But I never did.

I called Three Four again and let him know I was ready to register the battery. We proceeded to complete the registration mission. After the registration, I told the Bird Dog pilot we were free to start hunting targets of opportunity as we worked south along the Saigon River.

Sometime later, the pilot directed my attention to several men running in some trees. They were about fifty meters to the west of the bank of the river. There were four or five of them. They were running back and forth under the trees and it looked like they were shooting their weapons as they ran. They were Americans. It was clear to me that they were in contact and were trying to establish some sort of defensive position. I could not see what they were shooting at. None of them looked up.

"Ah, Ranger Three Four, this is Ranger One Two, over."

"One Two, this is Three Four, over."

"Three Four, One Two, it looks like I have some troops in contact, grid XT 685190, they look like they are in trouble, see if you can get all affirm air and ground. It will be a danger close mission, over."

"One Two, Three Four, can you identify the unit, over."

"Three Four, One Two, it appears to be a squad-sized unit in a defensive position. I cannot see any Charlie, over."

"One Two, Three Four, Wilco, over." I waited a couple minutes. The FDO came back on as we circled above the grid location. "One Two, Three Four, we have negative air and ground, over."

"Three Four, One Two, Roger, these guys need help, see if you can get a call sign and freq. (frequency) I will try to contact them, over." I thought we might be able to talk to them directly.

"One Two, Three Four, negative on the call sign, we checked again, negative on the air and ground, over."

"Three Four, One Two, it doesn't look good. I will remain on station for a while. Check again. If something isn't done soon, it will be too late."

"One Two, Three Four, Wilco, out."

We watched the action on the ground. We heard nothing through our radios. We just sat there in the Bird Dog, flying around in a tight circle about 500 meters off the ground. There was no doubt that men were dying on the ground below us. We were just a few feet away and we had the fire power to save them, but we couldn't. We couldn't even talk to them. It was as if we were in a movie theater; all we could do was watch, as spectators, their final, violent moments on Earth.

In my mind I could hear their screams for help. I could not stop watching.

The pilot flew the Bird Dog in a lazy circle over the men. Tears splashed against my sunglasses as I looked down on the carnage below. I understood what was happening, but I did not believe it.

A few moments ago, I thought, *things changed for those seven men. A few moments ago, I watched as their time ended. A few moments ago, I had the power to intercede. A few moments ago, I could have saved their lives.*

Maybe not, probably not, but now I'd never know. I should have tried harder. Maybe I could have been the hero instead of an impotent bystander.

I took a minute to clear off my glasses with some C-ration toilet paper I kept in my web gear. Once I had them cleared off and looked back down below, I was shocked. The scene was alive with Charlie. They were running all over the place, checking bodies and picking up equipment. My eyes zeroed in on one Charlie who had bent over and picked up an M-16. He held it over his head and began dancing around. As he did, he looked up at the Bird Dog.

Then he waved at me.

My first reaction was to grab my M-16 and start shooting. Then I remembered the radio.

"Ranger Three Four, this is Ranger One Two, over."

"One Two, this is Three Four, go, over."

"Three Four, this is One Two, you heard anything yet?"

"One Two, Three Four, negative, what is the status of the friendlies, over."

"Three Four, One Two, tell whoever is giving you clearances to contact the friendlies higher up to come and get them. Tell them to bring body bags. It is all over."

"One Two, Three Four, is there anything we can do? Over."

"Three Four, One Two, it's over, goddammnit. They don't need us anymore. They are all Kilo India Alpha, there are no prisoners. I count seven Kilo India Alpha. There appears to be a squad of Charlie gathering equipment, over."

"One Two, Three Four, I'll try again, over."

"Three Four, One Two, no need. I'm going home. Out."

Bien Hoa
Late Summer 1970

When we had finished our work for the day, SSG Son always had someplace for us to go before returning to Battalion. Thanks to the passes I had for the civic action projects, we were not prohibited from visiting areas that were off limits to most American military personnel.

We visited religious temples. We visited villages where the real Vietnam lived. We visited artists where the real art of Vietnam flourished. He was always careful to teach me about the history of Vietnam so, he hoped, I would begin to understand his country's needs.

I had mentioned to him that I wanted to get a doll for my daughter's birthday. But I wanted something that was authentic Vietnamese, I told him. Something not touched by America. He took me to the one place in the whole area that was strictly forbidden to me, the marketplace in Bien Hoa.

My first impression of the market was that everything was small. The canopies over the shops were built for Vietnamese, so I had to bend over all the time. I kept bumping against things, knocking thing over and causing a little good natured ruckus as I walked through the narrow aisles. I felt like I was in a dollhouse.

I saw gold, diamonds, C-rations, clothes, all kinds of meat, furniture, and little ceramic head-and -hands dolls made with traditional Vietnamese clothing. Everywhere I went, the people stopped and watched me. They knew I wasn't supposed to be there. But they were very friendly, smiled a lot and I had a great time.

In Vietnam, everything had several values, depending on which currency you wanted to use and what you were shopping for. There was one price for Vietnamese piasters, another for American military script and another for American Green Backs. There was a price for bartering with black market items such as C-Rations, M-16's or ammunition.

One of my favorite places to visit was an artist's colony set up by the South Vietnamese government. It was a safe haven for the artists on the road between Bien Hoa and Phu Loi.

We visited the colony several times. It reminded me of an art fair where you could walk around, watch artists work, and if you found something you liked, negotiate a price. At the front of the colony was a gallery where the best art from the colony was displayed for sale. Most of the art was metal work, painting, woodwork, and ceramics.

There was one painting that I really wanted. It was done on hand-rubbed teak with various shades of gray lacquer. It depicted a sampan, made of mother-of-pearl, egg shell and egg shell skin, on a foggy river in the morning. It was the most beautiful painting I had ever seen.

For whatever reason, I didn't buy the painting when I first saw it. I went back several times to look at it and for some reason could never pull the trigger on the purchase. Then it was gone.

As I remember, it was priced several hundred thousand in Vietnamese piaster, which was about $200 in military script (Military Payment Currency - MPC). The price was less for MPC, and even less in U.S. dollars. I had a lot of MPC on me all the time, Most everything I needed was provided by the military, so even though I sent most of my pay home, I still had more money than I could spend.

We had no access to U.S. dollars. In fact, it was against the law for us to have dollars. South Vietnam's "pi" was extremely depressed in value, so neither the American nor the South Vietnam government wanted dollars available to the economy. American dollars made it more difficult to control the value of the pi.

The reason we used MPC in a combat zone was to control who could use the money we brought in. The result was three different currencies: South Vietnam's, the U. S. military's, and the dollar. You could add the military currency used by the NVA and V.C. to make it even more confusing.

When we invaded Cambodia, we captured the occupation currency North Vietnam intended to use when they took over South Vietnam. I kept some as a souvenir and I carried it in my wallet until it rotted away.

On top of all that, you had the black market and the barter systems of currency. An M-16, for instance, would bring $700 MPC on the black market. That $700 would supply a soldier with drugs for a long time. And, when the time was right, that M-16 could be used to kill the soldier.

Eventually I found another painting. It was not a very good one, but it was real Vietnamese. It was also hand rubbed and inlaid with mother-of-pearl and that thin, paper-like substance found between an eggshell and an egg. I bought it and mounted it in my hooch.

After the unfortunate demise of John F Kennedy, the painting became my only personal connection with that room.

Soon it got around the battalion that I was getting to see Vietnam in a way that no one else was. I began to get requests from others to take them along when I went looking for art. I discovered that there were a few human beings left in the battalion after all. Suddenly, I was scheduling bus tours to Phu Cong for battalion soldiers. SSG Son acted as the tour guide to help them experience some real art and a different side of Vietnam than they were used to.

I justified the tours to LTC Mullens with an argument that some more good will might just help our war go by better and faster. I got the feeling he didn't need my argument.

—

Compared to that twilight zone of potheads, drunks, and Black Panther blacklists, chasing Charlie was fun. I became very cavalier about demonstrating my independence from the battalion and wore little badges to show I could get out of that pigpen. I wore the ARVN equivalent of my rank on my shirt, and carried my various special passes on my dog tags and in the fold of my sleeve. As a pièce de résistance, at least in my mind, I had my flight helmet painted with peace flowers. It was beautiful with all the reds, yellows, oranges and greens. The pilots liked it, too, since it made my head a better target than theirs.

The Battalion XO was a major, and that was also his name. Major Major. He had adopted the method of signing his correspondence as M^2. When he saw my helmet, he admired it and told me he wanted a poker hand with four deuces painted on the back of his helmet, with the caption, "Major Deuce."

For some reason it pissed me off that he would try to horn in on my little demonstration of rebellion. So I took his helmet to my painter and told him to do as the Major wanted, except that under the poker hand, he should write "Major Douche." The poor kid was afraid to do it—he was only a Spec. 4— but I assured him I wouldn't tell the Major who'd done it, and that I'd take all the flak.

The good major wore the helmet once. Somehow he was made aware of the wording on his helmet. When he brought the helmet back to me, I was sitting at my desk in the TOC. He slammed the helmet down on my desk and ordered me to fix it ASAP. Then he stormed out of the TOC.

CPT Miller, the Assistant S-3 on duty in the TOC, came over and asked what that was all about. So I showed him the helmet.

He burst out laughing. "Can I borrow this helmet for a few minutes?"

"No problem," I said. "Get it back to me as soon as you can."

Still smiling, CPT Miller took the helmet around to all the soldiers on duty in the TOC. For a few minutes, the atmosphere in the TOC was happy.

I had the lettering corrected. After I returned the helmet to MAJ Major, I was limping from the ass chewing I'd received. But it was worth it. The good major had a new call sign.

Best of all, no one else asked to get their helmets painted.

Most of the time I spent flying as an AO (air observer) was rather boring. My main mission as an AO was to register the firing batteries. A registration was a non-contact fire mission that could take considerable time to complete. These missions took up a lot of my flight time. The rest of my flight time was free to hunt Charlie.

One day we were flying south of Xuan Loc, in support of Charlie Battery and a small infantry operation being conducted by the Aussies and the 11th ACR. It was a Tuesday, the day after we took the big orange malaria pill. The side effects of taking that pill came the next day. Without exception, you'll find yourself paying attention to a serious case of diarrhea. There was no other option.

In the middle of a contact fire mission, I had a sudden, severe pain in my intestines, accompanied by an almost irresistible urge to shit.

Pill shit is like no other shit I have ever experienced. When it is time to pill shit, it is time to shit. If you don't shit, it feels like your whole insides are about to explode. It was that time.

My Bird Dog did not come equipped with a restroom. Nor could I hang my ass out of the window at two thousand feet and adjust artillery at the same time. So I rearranged my priorities. The fire mission did not have troops in contact, so its importance fell several notches.

I called the Charlie Battery FDO. "Ah, Ranger 44 this is Ranger One Two, check fire, over."

"One Two, this is 44, say again, over."

"44, this is One Two, I say again, check fire, over."

"One Two, this is 44, we have one in the tube, and check fire is not advisable. Please state nature of check fire, over."

"44, One Two, fire this round, do not reload, check fire caused by illness. I must land immediately, over."

"One Two, 44, Roger that. Is situation serious?"

That all depends on your definition of serious, asshole. "Ah, 44, this is One Two the situation is under control, I will contact you when I am airborne again. We will continue the fire mission then, over."

I spotted an old road beneath us in the jungle, and told the pilot he'd better land there or I'd shit all over his aircraft.

So he did.

As we started to land, the pilot radioed his people to tell them we were landing and give them our grid coordinates. We lost radio contact as we dipped beneath the jungle canopy. As soon as the

Bird Dog stopped moving, I jumped out, ran to the edge of the jungle, and did my thing. While I was in the process of finding some soft grass for the cleanup detail, along came a squad of infantry riding on top of an armored personnel carrier from the 11th ACR. No doubt it was too hot for them to ride inside.

They were pointing at the Bird Dog and talking. Finally, one of them saw me squatting in the bushes. He pointed. They looked. I waved, and they started laughing.

I guess it was better they were laughing and not shooting at me as I squatted in the grass. In the jungle, it could be very unhealthy when the infantry was around.

Then, as I was putting on my flight helmet, I heard:

"Rover craft, rover craft, this is Cowboy Six, commo check, over. Rover craft, rover craft, this is Cowboy Six, commo check, over."

"Cowboy Six, this is Rover Niner Seven, I read you, Lima Charlie, how me, over."

"Rover Niner Seven, this is Cowboy Six, I read you, Lima Charlie. My ETA to your location is five mikes. State the nature of your emergency, over."

"Cowboy, Rover, there is no emergency, we are okay. Passenger had diarrhea, didn't want him messing up my aircraft. I say again, there is no emergency, no assistance required, over."

"Ah, Rover, this is Cowboy Six, say again, over."

"Cowboy, this is Rover, I didn't want my passenger shitting all over my aircraft, over."

"Rover, this is Cowboy Six, I understand, break, Rover Control, Rover Control, this is Cowboy Six, over."

"Control, this is Cowboy Six, Rover craft is OK. There is no emergency. He has landed on a jungle road to let his passenger take a shit. I say again, there is no emergency, over, out."

Then, from some unknown station, came another broadcast. "You got to be shitting me, over."

I realized that when the pilot reported that we were going down in the middle of the jungle, everyone thought we had been shot down. So, as in all good standing orders, all airborne traffic was diverted to our assistance. Since we dropped below the canopy of the jungle before they could clarify the situation, they called it an emergency.

The net effect of the whole affair was that, for a short time, I had captured the hearts and minds of everyone in III Corps.

I had taken the most expensive shit in American history.

Battalion Compound
Early Fall 1970

One night, toward the end of my tour of duty in Vietnam, I was serving as OD (Officer of the Day) and was making rounds of the perimeter, expecting nothing to happen. Since the flash base tower was close to the TOC (Tactical Operations Center), I usually checked it last. It was one of several strategically located towers in the Long Binh, Ben Hoa military complex. Its purpose was to find the location of rocket and mortar positions by triangulating with the other flash base towers.

I did not like to climb the ladder to the top, so I only did so when it was absolutely necessary. The tower swayed as you moved up the ladder. The rungs were made out of metal, and they were slick. I had an aversion to long ladders; the further I got from the ground, the less I liked them.

I had developed the habit of checking the tower once during each tour as duty officer. The rest of the time, I'd just step on the bottom rung and the vibrations would alert the guards up in the tower. They would look over the edge to see who it was, and thus I could verify everything was OK without actually climbing all the way up.

The reason this worked so well was that the guards were afraid of the tower, too. They knew Charlie knew what the tower was used for, and that it was on the top of Charlie's target list. The guards were afraid that a sapper could catch them unaware. As a result, no one climbed that ladder without being challenged.

This night I was short, with about 30 days left before I would rotate out of Vietnam back to home. I had a short timer's attitude, which meant I was losing interest in my job and getting a little lazy. I decided I was not even going to climb the tower tonight. I even entertained the thought of passing it by. As I walked by the tower, I heard loud music coming from the top. I changed my mind, walked to the base of the tower, and stepped onto the first rung of the ladder, still not intending to go all the way to the top.

Nothing happened. The music continued. So I stepped up to the second rung, then the third. Nothing happened. No one challenged me. So I kept moving up the ladder. Halfway up and there was no challenge, so I kept going. When I reached the top, I looked over the steel plating used to protect the guards from small arms fire. I didn't see anybody.

The platform at the top of the tower was about four meters by four meters in size. There was just enough room for the BC (Battery Commander's) scope, personal gear, a couple of chairs, and a cot for the off-duty guard. We used the BC scope to establish an azimuth (direction) to the flash of the mortar tube or rocket launcher. The BC scope was the most critical piece of equipment in the tower.

I noticed no one was manning the BC scope. On further inspection, I noticed that a tent made of ponchos had been erected in part of the tower platform. As I started to crawl over the steel-plating wall, a head stuck out from under the tent and looked me straight in the eye. The head ducked back into the tent, and I heard:

"Oh shit, it's an officer!"

The tent burst open with bodies scrambling out and standing up. All of a sudden I found myself face to face with five soldiers, one white and four black.

I demanded, "What the fuck is going on here?"

One of the soldiers spoke up. "We're just rapping and listening to some music."

"Who the hell is supposed to be on duty here?"

Three soldiers identified themselves. I remembered them from guard mount. I didn't know anything about the other two.

"Who gave you permission to erect this tent up here?"

One of the guards who was supposed to be on duty responded, "No one, sir."

My first thought was to tell them to tear down the tent and to order the two extra soldiers to leave. As I talked to them more, I realized that none of them were really responding to anything I said. So I decided to have a closer look. I got into the tent and found seeds all over the floor. I asked:

"What are these seeds?"

One of the soldiers that was not supposed to be in the tower said, with just a little bit of defiance in his voice; "You know what they are, sir."

I really didn't, until that moment. I inspected all their personal equipment and found several bags of grass. I had no idea what the seeds were.

When I was satisfied that I had found everything I was going to find, I grabbed the telephone to the commo room and gave the ringer a spin.

A voice came on. "This is the perimeter switch."

"This is CPT Martin, the OD, speaking. I am in the tower. Get me the NCOIC, ASAP."

"Hold on, Sir."

About twenty seconds later, the non-commissioned officer in charge came on the line. "This is SGT Finch speaking."

"SGT Finch, this is CPT Martin speaking. I am in the flash base tower. I've just caught the guards and two other troops having a pot party up here. I am relieving them from duty. I want you to get me three more guards who know how to run the BC scope up here on the double. I will stay here until they arrive."

"Yes, Sir."

I hung up the receiver and turned to the troops. They were standing in a semi-circle around me, just out of arm's length.

"OK, gentlemen," I said, "you are relieved of duty, You can leave the tower now."

No one moved.

"Gentlemen, it appears you misunderstood me. I am ordering you to leave the tower."

Nothing happened.

Well, my friend, you stepped in it this time.

I backed away from them until my spine touched a steel beam in the corner of the platform. I slowly reached up and took my 45 cal. pistol out of my shoulder holster. I wanted them to see I was serious. I jacked a round into the chamber and pointed the pistol toward the sky.

"Gentlemen, I don't know what you got in mind, but I can tell you this. If you don't start moving, it is going to get very messy, very quickly. You may get me, but I promise you, some of you won't live to tell about it."

A cold wave of fear hit me. Was I going to be one of those statistics nobody wanted to talk about? Just an unfortunate officer who took the wrong way down from a sixty-foot tower? It was a perfect setup; all they had to say was that I slipped crawling down. Who would know? There would be no witnesses.

I thought the only thing saving me was that call to the commo room.

After a moment, they started to move. One by one, they went over the side and started down the ladder. As I watched them crawl onto the ladder and start down, I realized I knew these guys. It hit me like a bolt of lightning. Three of them were part of the infrastructure of the Black Panthers. I didn't know anything about the fourth one.

My friend, what have you got yourself into now?

The last one to leave, a Black Panther, paused as he climbed over the side and turned to me.

"Sir, are you going to report us?"

"I think I will, yes."

"You will be sorry if you do—Sir." Then his head disappeared below the steel plating on the side of the tower.

I watched them descend the ladder to the ground. The tower stopped vibrating as the man who'd threatened me stepped off the last rung. They disappeared into the night. I turned to look toward the northern field of observation. I remembered I still had my pistol out. My hand shook as I ejected the magazine, ejected the round in the chamber, replaced the round into the magazine, and replaced the magazine into the pistol. Then I returned it to its holster.

In a whisper, I said to myself, "I hate this goddamned tower."

It took probably thirty minutes for the NCOIC to find three guards who knew how to operate the BC scope. The time crawled by. The thirty minutes seemed like three hours. As I waited, the tower began to take on the proportions of that bunker in Tay Ninh. It grew larger. The black night in the tower was almost as black as the black in that bunker. I started having trouble breathing. I relived that feeling of the rat on my chest. Then I could feel the cockroach crawling on my stomach toward my chest.

The next day, after some back-and-forth mental fighting between being short and not wanting to rock the boat and being an officer knowing what my duty was, I decided to report the incident. I called the military police operations center in Long Binh, identified myself, and asked for the officer in charge.

"Wait one, Sir."

After a short pause, another voice came on the line. "This is Lieutenant Whit, OIC. How can I be of service?"

"I am the security officer for Camp Price, the compound just west of Widow's Village," I said. "Do you know where my compound is located?"

"Yes, Sir."

"Last night we had an incident involving a pot party. I had to confiscate several bags of pot that I have since secured in a safe. I need your direction on how to dispose of it."

"Captain, we won't accept it, and if we catch you with it, we will arrest you."

"What?"

"Sounds a little weird, sir, I know. My best advice to you is to get rid of it ASAP."

What the fuck, over. "Yeah, this is weird," I said. "Thanks, but no thanks." I hung up the receiver and wondered what to do next.

I arranged to have the pot destroyed in a big fire on the parade ground. A couple of days later, I learned through the grapevine that I'd been blacklisted—targeted for elimination by the Black Panthers.

There were three names on the blacklist. Mine, an NCO's, and another captain's. The NCO's name was on it because he liked to take dope away from the troops and make a big production out of destroying it. The other captain's name was on it because he broke up a riot one night by threatening to blow the head off of one of the brothers with a sawed-off shotgun. The riot had started when several black soldiers cornered some whites who happened to be on guard duty. The captain, who was the OD that night, was alerted to the situation and he ran to the location of the confrontation. The black soldiers had surrounded the white soldiers, but the situation had not escalated to physical violence yet.

He waded into the riot with his sawed-off shotgun—a nasty little thing with a nasty reputation. He found the center of gravity, picked out one of the Black Panther leaders, and jammed the shotgun's barrel up under his chin. Then he said that if things didn't return to normal quickly, this brother would lose his head.

For my remaining time in Vietnam, I actually forgot about Charlie after the sun went down. Charlie wasn't the reason I slept with a loaded pistol in my hand.

I was real happy that my bunk was next to Doc Carr's bunk in the hooch. I figured they wouldn't frag me there. Most of the rest of the time I was not in the compound, so it would be hard for them to set up an ambush.

Long Bien
September 1970

Short timer's fever did not get to me until the last couple days of my tour. Until that last week, I stayed busy flying and working with the guys in gray.

Over lunch with them one day at the Long Binh II FF officers' Club, they asked me if I'd consider extending to work directly for them. My job would be to coordinate fire missions for Army, Navy and Air Force assets on their targets. I asked if they could guarantee that I wouldn't have to deal with dopers. They wouldn't make that commitment, so I said no thanks.

I sometimes wonder if I made the right decision. There are times I think this part of my war might have been more interesting had I chosen to work with the guys in gray. Then I have a sober moment and remember all that has happened because I didn't.

After incidents with dopers and Black Panthers, I'd lost my fear of Charlie lurking out in the jungle. Instead, I was afraid of everyone else inside the wire with me. I decided to find as many excuses as possible to stay outside the wire, where it was safe.

Short timer's fever got everyone in Vietnam. It got you when you realized you didn't have much time remaining before you'd rotate out of Vietnam and go home. It is a kind of insidious condition that changes a get-it-done attitude to a let-someone-else-do-it attitude. It comes from the realization that you may just live through your experience in Vietnam. So living becomes real important to you.

At least that is how short timer's fever got to me.

I can remember only one incident where I can say that short timer's fever caused me to overreact. The day before I was to report to the replacement battalion, LTC Mullens called me in and ordered me to make a solatium payment to a farmer who had some fruit trees damaged by one of our artillery rounds. He gave me the grid location and I went back to my office to determine where the farmer lived.

It was the same grid location as that of a VC regimental headquarters.

I couldn't believe it. I was set to rotate home the next day.

Before I could gain control of my mouth, I blurted out, "God*damn*, that son of a bitch wants me to go out into the middle of a VC regiment and pay some goddamned farmer for some goddamned trees that lost some goddamned bark. Why can't some other goddamned fool do it?"

I was still standing at the intel map. The whole operations section turned to look at me. "Well, shit, Sir," I said to MAJ Brown, the operations officer. "This ain't fair. I'm too short for this kind of bullshit."

I turned to my driver. "Smith, go get the goddamned Jeep, find SSG Son, and report back here as soon as possible."

Then I continued my tirade.

"What's so fucking important about a couple of trees?" I was holding the pointer I had used to determine the location of the damaged trees. I threw it at my desk.

"I don't give a damn one way or the other about these goddamned trees. They probably belong to the goddamned VC regimental commander." I stomped over to my desk and picked the pointer up off the floor. "In three days, I will be a civilian and I won't have to put up with any more of this bullshit. What a fucked-up organization. We are now paying our enemy for shooting at him. No wonder everyone is so doped up around here all the time. Jesus Christ."

MAJ Brown said, "That's enough, Captain."

"Yes, sir."

Actually, the trip turned out to be very enjoyable. We drove along the coast toward Vung Tau, stopped and chatted with some Australian friends for a while, then drove north toward Zuan Loc. We had no trouble finding the farmer. He was no VC regiment commander. He was just a farmer. He took the envelope with the solatium payment and gave SSG Son and me some fruit to eat on our way back to Battalion.

I got to see some of Vietnam that I'd flown over many times, but had never seen up close. It was a pretty part of Vietnam. There was jungle, interspersed with manicured fields and rubber trees. White clouds floated in a blue sky, and the wind caused by the Jeep as it traveled down the road kept me reasonably comfortable, temperature wise.

Vietnam was a beautiful country, and the Vietnamese people were nice to be around.

—

During those last days, I thought a lot about extending my tour, provided I could get the guys in gray to have me reassigned. I kind of liked the intel business. But in my saner moments, I realized there was no way I could escape the drug problems in Vietnam. So I dismissed the idea of staying longer than I had to.

However, as a last demonstration of my independence, I organized a dope raid on the battalion. It took me weeks to set it up. I had discovered that if I got any unannounced inspection or raid approved by the command system, somehow the dopers always knew we were coming. As a last great farewell to my wonderful world of night crawlers, I contacted an MP unit with sniffer dogs and arranged to have the complete compound cut off while they made a sweep.

I was so concerned about leaks that I didn't even use my usual driver when we met to organize the sweep. The MP unit assumed I had approval from my commander. Apparently they didn't think a captain would show that kind of initiative. They were wrong.

The sweep was to begin thirty minutes after I signed out of the battalion roster on my way to the repo depot, the replacement battalion for incoming and outgoing military personnel. I expected the operation to produce a lot of dope. Maybe my skullduggery would do some good, I thought. But my real purpose was nothing more than revenge.

My plan didn't work out. The morning I was to leave for the repo depot, MAJ Major called me in and asked me to explain the phone call he had just received about a drug raid later that the day. I told him what I had arranged and explained why I felt we needed the raid. He then told me that the raid had been canceled due to another mission of higher priority.

But he agreed that a raid was a good idea. He said I should set it up again, but this time I was to work with him. He was stern when he said it, making sure I knew he was serious.

At this point I was so short I didn't even have a wakeup between me and my freedom bird. I suppose I should have been disappointed, but I wasn't. I didn't care one way or the other. I didn't have the guts to tell him I felt that he was part of the problem. The real reason I had set up the raid was to catch him or anyone else who might be corrupting the chain of command. I told him I'd reschedule the raid, even though I was already packed and my transportation was on the way.

I smiled as I left his office, thinking *Don't hold your breath on that one.*

My next and last act as an officer assigned to the battalion was to sign out. I was one happy son of a bitch when I climbed into the Jeep and left, never to return.

Well, asshole, it looks like you're going to survive this nightmare. But it ain't over. Not by a long shot."

As I rode in the Jeep traveling toward the Bien Hoa Air Base, SSG Son popped into my mind. I was going to miss hearing him say, "Numma fucking ten, *Diwi*."

That ruined my mood.

Freedom Bird
September 8, 1970

There was no thrill in leaving Vietnam, not like I'd anticipated. Far from a feeling of relief, I was feeling more stress, just a different kind of stress. I was heading back to America, a place I had left three years ago. A place where people lived who did not like me because of the uniform I wore. I was going to touch my daughter again. I was going to have to try to live with my wife. My parents would help me.

It made me angry to see women protesting the war in news clips on the TV. Kent State still pissed me off. Privileged characters like college students and their pampered professors were calling people like me baby killers. These people were throwing rotten food and dog shit at people like me. Sometimes they would even kill people like me. I could not understand why they thought they had the right to judge me. Where was their pound of flesh?

When I saw college students protesting the war, I could not understand what they had to protest about. They were safe, clean and dry. Most of them would never serve their country in peacetime, little alone war. Most of them had nothing to fear about this war. It was not their war.

My only contact with the people of America that were not in a uniform for three years had been through TV. Television made it seem like nothing I had done in Vietnam had value. Maybe Mr. Son was right; maybe I should question if America was there for me. I did not know the people in the news; I did not understand them and I was afraid of them.

Now that I was about to climb aboard the freedom bird, I became confused. I knew I needed help. But on the TV screen that had become the face of my America, I saw no one I could trust. Maybe, I thought, it would be best for me to stay in Vietnam. At least I understood the risk of living or dying there better than I did in America.

Checking our baggage at Bien Hoa Air Base before we boarded, we were told that if we had any weapons, we were to give them up. Give up our personal protection while we are still in Vietnam? *I don't think so,* I thought. My Chicom (Chinese

Communist) K-54 pistol was in my briefcase, and it was going to stay in there, as far as I was concerned.

All of the soldiers scheduled on the flight were led to the briefing room. The briefing officer began by explaining several forms we were supposed to fill out. Then he told us that several soldiers had been killed in San Francisco within the last month. He warned us that it would be best for us not to go into San Francisco at all, but if we did, we should not wear our uniforms. Then he told us to open our briefcases. He started down the aisles inspecting the contents.

When he came to me and saw my pistol in my open briefcase, he looked up at me. "Captain, you can't take that pistol on the plane with you," he said. "You can take it home if you'll agree to check your briefcase into the luggage compartment. Otherwise it stays here."

So the briefcase rode home with the luggage in the belly of the plane. For the first time in months, I was without protection. I was still in Vietnam, but I could not defend myself. Very unsettling.

—

The airplane's speakers switched on and a man's voice filled the cabin. "This is your Captain speaking. We will be coming up on the Californian coast in about ten minutes. We will be landing at Travis in about forty-five minutes. Welcome home." As I watched the coast of California slide by, tears came to my eyes. It didn't feel right. I was supposed to be dead. I hadn't prepared for living.

I don't remember landing or getting off the plane. I do remember walking down a long hall with a big sign at the end that said, "Welcome Home, Job Well Done." I remember boarding a bus with my discharge papers in hand. The bus took us to the San Francisco airport. I was still in uniform, so my senses were keen when I got off the bus. It was no time to fuck up.

I entered the terminal and made my way to the United counter. The woman who greeted me was the most beautiful creature I had ever seen. She had long, dark-brown hair, perfect skin, and beautiful blue eyes.

I placed my baggage in the opening at the counter, gave her a copy of my orders, and told her I wanted a ticket to Salt Lake City. I felt myself smiling at her. As she was making out the ticket, I said, "I have a war trophy in my briefcase, what do I have to do to get it on the plane?"

"What is it?"

"It's a Chinese communist service pistol."

"Why didn't you leave that filthy thing in Vietnam?" Her tone was cold and biting. "We shouldn't allow people to have things like that. You can check your briefcase if you want."

I had not expected such a response. As I stood there staring at her, I swear, her features began to change. She started to look like a rat. She had lowered her head, like she didn't want to look at me.

Another United clerk, this one a man, stepped next to her. He looked at her, then at me. "Is anything wrong?" he said.

"Ah, no," I said. "Everything is okay. Thank you."

Be careful, my friend. They may have been right about San Francisco.

The World
September, 1970

It was dark when the plane landed at Salt Lake City. The excitement was building in me so fast, I wanted to push everyone out of the way. My wife and my daughter would be waiting for me at the gate. I couldn't wait to hold them. I had hand-carried Nurse Quan's tiny tea set all the way from the clinic and desperately wanted to give it to my daughter. I had nothing else in my hands.

As I walked through the door, I saw another soldier get rushed by his family. There was a lot of noise and tears. His mother and father were crying, his wife or girlfriend was crying. They just closed in around him as if to welcome him back into their hearts. The excitement was overpowering. I looked around for a familiar face.

But there was no one.

My wife was not there. My daughter was not there.

I stood there several minutes. The crowd thinned out until I was the only one there. I didn't know what to do.

My friend, you have been forsaken. By your country, by your Army, and now by your family. What do you do now?

For several minutes, I stared down that long air terminal corridor. There were no other flights arriving or leaving. The corridor was empty except for me. The tears began to flow.

This, my friend, is what you get for not holding up your end of the bargain. Your job was to die so that your family could live. How selfish of you.

I trudged mechanically toward the main terminal. I knew I would have to claim my baggage, but I had no idea what to do after that. Should I make a phone call? She knew that I would be on this flight. Should I get a cab and go into Salt Lake City? Should I get another ticket and fly on to some other place?

I was waiting at the baggage carousel when I felt a hand on my arm. I jerked my arm away. My wife was standing there with my daughter, Anna, in her arms. She was sorry, she said, but the traffic had held her up.

She handed Anna to me, but as I tried to hug her, she pushed away. My daughter did not recognize me, or maybe she was afraid of me. Maybe I should have anticipated this reaction. My wife had made it clear she didn't feel comfortable around me in her letter several months earlier. It kind of made sense that my daughter would feel the same way.

I gave the tea set to Anna. She took it, looked at it, and lost interest in it. My wife did not ask me what it represented to me, so I said nothing more about it. If there were hopes and desires for help or understanding inside of me, they began to shut down. I never felt so alone.

Later, I found myself in a motel room with my wife. I sensed that she did not want me to touch her. That hurt. I had kept my marriage vows; I was clean. I hadn't brought home any diseases. I remember her asking me if I was still in the Army. When I said no, she got hostile and demanded to know how I planned to support her.

All I wanted to do was have her hold me. I didn't want to think about the future or the past. I just wanted to be loved.

Listen, my friend, your war has changed now. The bullets you dodged in Vietnam were just body busters. The bullets you face now are more vile, contemptuous, and infinitely more dangerous than anything back in Vietnam. These new bullets are designed as will busters.

Take care, my friend, you must protect yourself or you will be lost forever.

It would be a long time before I would ever ask for help again. To my utter amazement, my war was not over. It had simply taken on a new dimension.

Rob
December 1967

Home. It had been winter when I left for Vietnam. When I came back, it was fall. Other than that, it appeared nothing much had changed. I was an officer and a gentleman in the winter, and a nothing in the fall. I had done a lot of changing, so participating at home in the fall did not come easily. There are days, weeks and months at the end of 1970 that are still lost to me. My memories from that time are like snapshots glimpsed through a perpetual fog.

There are some good memories of that time. There is the memory of Anna. Within a short time, she accepted me as her father and her acceptance was total. There is the memory of my parents. They tried to make home the home their son had known. I had left that son somewhere in Vietnam, but they were happy with the person who came back. And they were not judgmental. They loved me as they had always loved me, as their little Johnny. I had neither the will nor the ability to tell them Johnny had died in a deep dark bunker in Vietnam. I let them love their son as if he were me.

And there were my friends Rob and Reggie, who loved me as damaged warriors love one another. They listened to my silent cries, and they understood. There are memories of friends, veterans themselves, who were ahead of me in the struggle to return home. They couldn't walk the road for me, but, they could alert me to the problems I was about to experience.

It was time for me to learn to love again.

I am sure there were times when it was not so good being home. It is these memories that are lost in that perpetual fog. I am okay with these memories staying lost.

In 1967, before I ever set foot in Southeast Asia, my friend Rob had already returned from Vietnam. He was in the infantry. He brought home the dreams, the residue of his violence, some he made and some that were forced on him. As luck would have it, we were able to spend our leave together before he returned to his stateside assignment and I left for Europe. It was a two-week party full of booze and broads, and if there were no broads available, we lived in the whorehouses.

Rob's dreams took a part of him away from home. I was shocked at what the dreams did to him. Something in Rob's brain was broken.

One night he was driving his brand-new car and he parked it in front of a bar. Our goal was to get shit faced. He shut the engine of the car down, turned to look at me in the passenger seat, and broke down. He started talking about the terror of being ambushed in an armored personnel carrier (APC), being cut off, wounded, and alone. He was the only survivor of the initial ambush. The APC had been disabled with an armor piercing grenade. Somehow his squad and APC had gotten separated from the rest of the platoon.

After the explosion, Rob recovered his senses and realized he was in deep trouble. He tried the radio and it worked. The platoon leader responded, telling Rob they could not get to him. He was on his own. At that moment, Rob began hating officers. The company first sergeant called him on the radio and started talking to him, telling him to settle down and take charge of his situation. He had Rob move to the .50 caliber machine gun and begin laying down covering fire. As Rob responded to the first sergeant, he created the time needed for his rescue.

We sat in the car for a long time while Rob recovered. I vowed that would not happen to me. I vowed that if I went to Vietnam, I would die before I came home as only a part of a man. I would not be mindless.

In the fall of 1970 I was wrong. I didn't keep that vow.

—

About a week after I got home, I was staying at my parents' house with my little family. We slept in my old room in my old bed.

We had sex in the dutiful way that partners do when they submit rather than give and take. It was an act, purely physical. We went to sleep without really touching each other physically or emotionally. We did not say good night to one another. It was just as well. Once we slept, we were free of each other for a few precious hours.

I soon found out that I had brought my own dreams home, just like Rob. They were not as easy to shed as my uniform.

I had been in a deep sleep when I sensed someone was looking at me. My body tensed as I slowly opened my eyes. I was aware of my wife, in bed with me and of Anna in her crib beside our bed. I struggled to continue to breathe normally. My heart was pounding. Slowly, I looked around the room looking for something that didn't fit. Then, as I looked toward the door I saw the dead NVA soldier, the one who'd been on top of the pile in Cu Chi, standing there with his eyes on fire, smiling at me.

I rolled out of bed and crouched behind a desk. After a moment, I looked around the desk. I could see his burning eye. I might have been dreaming up until that moment, but from then on I was awake, or at least it seemed that I was.

For a few moments I tried to decide what to do. I had to protect my family.

I jumped up and dove over the desk. As I did, I kicked a wastepaper basket and sent it sprawling. As I lunged at him I kicked a chair and it rolled away from the desk. I reached out and grabbed for his eyes. It was time to kill this man.

And then I really woke up. There was no man. There was a light switch with a light in the handle.

I felt like an idiot.

So I switched on the light. I switched it off and then on again. There I stood holding onto the light switch with no clothes on. The chair was on its back laying against the wall. The waste paper basket was upside down against the other wall. A dirty diaper was on the floor next to my left foot. I had torn the room apart. My wife sat up in bed. She screamed at me. "What are you doing?"

What was I going to tell her, that I'd just tried to kill a light switch? "Nothing," I said. "Go back to sleep."

I turned off the light, walked back to the bed, and crawled under the covers. I rolled away from my wife and went to sleep. The last thing I heard that night was my wife saying, "I want my mother."

She and I tried, for our daughter's sake, to find common ground from which we could build our marriage. While I had been in Vietnam, she had received her college degree and was trying to start a career. In November, I found a job as a common laborer in a construction company. I was making 75 cents an hour. Our combined income was enough to pay for rent for a one bedroom dump, basic food, and nothing more. We were existing as a family at poverty level, but we were not living as a family. The comfortable life as an officer and an officer's wife was a thing of the past.

In those first months, I was trying to find something to grab onto that would give me purpose. I was unable to sustain a nurturing relationship with my wife because I felt, down deep, that what had happened to me had been her fault. I can now acknowledge it was irrational, but at that time in my life, I was bitter that my wife, and women in general, would never have to suffer the reality of war. I resented the fact that she used the blood money I earned while I was in Vietnam to enrich her life. I resented the fact that I was trying to kill light switches and she didn't seem to care.

The first casualty of the second phase of my little war was my marriage. Other than my relationship with Anna, I saw no value in it. By an unspoken mutual agreement, we decided the only real purpose to our marriage was to maintain a safe environment for our daughter. I had spent several months in Vietnam living in an emotional vacuum my wife wanted. I could see no reason that vacuum could not continue. For Anna's sake, we reached an unspoken agreement. For the immediate future, we would not confront each other on substantive interpersonal matters.

Wyoming, 1970
Hunting Season

I grew up in the mountains and deserts of Wyoming. The wind, the sun, the earth, and the water, the sense of freedom, and the appreciation of the natural beauty were in my soul. My parents had made it so as I grew from a baby to manhood there. It was only natural that they would assume that Wyoming would help me recover.

Hunting season had always been my favorite time of the year. It was hunting season when I got home from Vietnam. So my dad took me antelope hunting. I think it was important to him that he and I spend some time alone together. I think he wanted to find out how much damage had been done. We could find that private time while hunting.

It was good to be back. It was good to be out in the country I knew so well. It was a familiar and friendly place. I knew my dad wanted to be with me and I knew he would be always be there if I needed him. I couldn't admit that I was damaged even though I knew I was. My dad wouldn't talk to me about being damaged, even though he sensed I was. At a minimum, I was mad and he knew it.

I wanted to be in the country and I wanted to be with my dad, but I did not want to hunt. I couldn't tell him that; he was trying too hard to do the right thing. He knew me as a hunter. I would do as he wanted, I would hunt and maybe kill again. For him. He was a good man.

I knew I was lucky.

As we traveled across the open country in an old green Ford pickup, looking for antelope, I held my rifle in my hand, rested my arm in the open passenger side window, and enjoyed the fresh, clean air in my face. The sky was blue with a few small clouds floating from west to east. The prairie was brown from the effects of the fall season. From horizon to horizon, nature provided me a beautiful panoramic view full of all the forms of life unspoiled by the greed of man.

It was good to be back.

After a time, my mind wandered and I began to daydream. As I scanned the prairie on my side of the vehicle, I felt myself being transported slightly above the terrain. I would catch myself looking for another kind of target. I felt the sweat on my hands as they held not my .30-06, but an M-16. It felt so real. I was back in Vietnam, in a Jeep looking for someone out there about to shoot at me.

My dad said something in a conversational tone. I guess he thought I had been quiet too long, and I snapped back into reality. "What?'

"I was wondering if you are still in the Army."

"No, I've been discharged. I'm a civilian now."

"No more flying off to strange places?"

"No, I'm done being stupid."

Dad chuckled: "We can only hope, but I won't hold my breath on that one."

Late in the afternoon we stopped on a bluff overlooking the large, flat valley we had just crossed. We talked about nothing important as we made sandwiches and drank coffee. My sandwich was a work of art: rye bread, about one inch of ham, two slices of Swiss cheese, mustard and mayo, two thick slices of tomato, lettuce and pickles, and another piece of rye bread. The only way I was going to get it into my mouth was to sit on it and squash it flat. I took my paper plate with the sandwich and some chips and my cup of coffee and moved away from the pickup to sit on a ledge of rock that overlooked the valley.

The day was warm but not hot, and the sun was bright but not overpowering. There was no wind blowing. I planned to enjoy my sandwich.

I took my first big bite of the sandwich. I had to open my mouth so wide I thought I'd break my jaw. It was a good bite.

I chewed a few times, swallowed, and just as I reached for the cup of coffee, the sound of two shots washed over me.

They were a long way off. There was no echo. Whoever was doing the shooting was shooting in our direction. I felt danger. I picked up my sandwich and coffee and moved behind the pickup for protection. Dad was sitting on the tailgate, eating his sandwich. I told him someone was shooting at us.

Then I corrected myself. "Well, maybe not at us, but in our direction."

I knew they were shooting at antelope, but it made no difference. I wanted to hide. As I stood there on the protected side of the pickup, I noticed some antelope run under the bluff and circle back to the right.

Dad said, "Look there," pointing to the antelope.

My first reaction was to watch the antelope run. At a distance, antelope are beautiful animals of the goat variety. They are not so pretty up close. They are smelly and their coat is more like feathers than fur. Up close, they look mangy.

When antelope run, they are fast and magnificent. I was not going to do anything about it, but my dad got excited. But his rifle was on the gun rack in the cab of the pickup, unloaded. There was no way for him to take a shot.

So I put my sandwich down, grabbed my .30-06, and ran to a point where I could get a better look and a better shot. There were a couple of nice bucks in the bunch. I pulled down on the first buck.

Look at that, my friend. Behind the buck is a round eye. You're finally going to get your shot at that bitch.

I switched my target from the buck to the doe and pulled the trigger. She went down like a rock. I saw the bullet strike her just behind the front shoulder and I saw a spray of blood into the air on her other side where the bullet exited her body. I continued to look through the scope as the sensation of the explosion of the rifle subsided in my body. The doe was dead before she hit the ground.

I should have been upset about shooting the doe, but I wasn't. That it was so easy to kill scared the hell out of me. Hadn't I learned my lesson with the dog?

Dad was still standing by the pickup eating his sandwich as I walked back to get the knife and hatchet to dress the doe. I told him that a doe had jumped in front of the buck just as I pulled the trigger.

"That's okay, Son. Doe will probably be better eating anyway."

I felt a strange sensation in my stomach. Was it pain? Was it excitement? Was it tension? I couldn't tell.

I didn't like it.

—

My mom's simple gesture of love for me was cooking the foods that her Johnny had once liked. I was still a steak and potatoes kind of guy, but if mom sat a plate of Swiss steak with some new potatoes and corn on the cob in front of me, I would pass up a T-bone any day. I was a cake and milk guy, fill the glass up with cake, pour on the milk, and eat it with a spoon, but if mom sat a pan of bread pudding in front of me, I'd pass up that cake in an instant.

That was something that Vietnam did not change. As her food touched my lips and passed into my body, I knew she'd prepared something special for me. This was her way of protecting my soul.

I came to understand that mom and dad were having a hard time dealing with the stranger I'd brought home to them. They tried to talk to their Johnny several times those first few months. Then, as time went by, they began talking more and more to John.

Rob, Reggie and I were good friends who had grown up in Rawlins. We spent a lot of time in the country and had developed a kinship with the personal freedom that was unique to Wyoming. Our overactive imaginations as young adults allowed us to experience America as we thought our ancestors had. We could be a part of the Wild West as we traveled and investigated the relics of times past. As we explored the old homesteads, the Indian campsites, the old mines and the logging camps, we could easily imagine that we were the men who had defined the Wyoming legacy.

By the time I came home, Rob was an old hand at the second phase of the war, the little war. He had spent three years fighting his little war—the dreams and social discrimination that resulted from the contamination of Vietnam veterans represented in peace-loving America. He was used to that "get over it and get on with your life" attitude he was confronted with any time he tried to talk about Vietnam. No one wanted to hear about his emotional pain. So he did what the other vets did. He shut up about Vietnam when he was in public. And the pain was tearing him apart.

He was there when I got home.

Reggie was in the 11th Armored Cavalry Regiment in Vietnam at the same time I was flying observer missions in support of their ground operations. It was very possible I fired artillery missions in support of Reggie. We had no idea we were that close together until after we got home and talked to each other.

For a while, Rob, Reggie and I worked as laborers in a construction company. During our breaks and lunch we would sit down and talk about Vietnam. Some deep, dark secrets came to the surface for each of us, and we talked to and mentored each other. Then one day, during one of those deep conversations that only we could understand, one of the carpenters came over to us and told us to shut up.

So we did, that was the end of that.

Reggie was killed in a car accident shortly after he got home. I couldn't attend his funeral. My emotions would not let me.

I should have attended his funeral.

One day in late September, Rob and I were returning home from a trip to the mountains south of Rawlins. The weather was clear, the temperature was shirt-sleeve warm, and the air was fresh. It had been a great day with just a hint of that old-time feeling of freedom we used to get as young boys, pre-Vietnam.

We were traveling on a dirt road with dust boiling up behind the pickup. Rob was driving and I was daydreaming in the passenger seat. We topped a hill came upon a badger crossing the road from left to right, right in front of us. Rob slammed on the brakes and the truck slid to a stop.

Rob handed me his pistol. "Shoot it."

The badger was rambling away from us up a hill, no more than twenty feet away. I jacked a round into the chamber and pulled down on it, bringing the sights in line on the upper back just behind its head. The badger stopped, turned its head, and looked right at me. I flashed back to the dog. The dog that had trusted me, the dog that I'd killed as it watched me, expecting me to show it mercy.

All I could see was those big brown eyes.

I fired the entire magazine to the right of the badger. I wanted to start him moving so Rob would not get a chance to shoot at him. After I fired the last round I watched the badger disappear over the hill.

Not today, I thought. *Killing can't be a part of this day.*

Beside me, Rob laughed out loud. "God damn, John," he said. "You missed him every shot." But as I handed him his pistol and he looked me in the eye, his expression changed. When he spoke again, the laughter had gone out of his voice. "Oh, I'm sorry." he said. "I forgot."

Soon it was deer season. Rob lined up a hunting trip at a ranch south of town. Dad, my Uncle Darold, Rob and his dad, Pat, and the ranch foreman were our hunting party. We set up camp back in high country, about thirty miles south of Rawlins on the edge of the Medicine Bow National Forest the night before hunting season opened. Uncle Darold had brought a small camping trailer for us to cook and sleep in. We were roughing it in style.

It snowed about three inches our first night in camp. We rousted out of our fart sacks about four a.m. and started the morning chores. I walked out into the open to take in a deep breath of the cold, clean air and kick at the soft, fresh snow. I stretched my arms, straightened my back, and looked to the heavens. The stars were there from horizon to horizon, north to south, east to west, bright and clear against the black sky. It was going to be a perfect day for hunting deer.

We were up and out of camp before daylight. We drove our pickups to the open end of a box canyon with high ridges on three sides. The ridges were covered with buck brush, a perfect place for deer. As the sun broke over the east ridge, while we were still in our pickups, Uncle Darold saw deer moving in the brush under the ridge to the south, about a thousand yards away and two hundred feet above us.

Dad had a bad foot and could not easily walk in broken country. Rob's dad, Pat, had some health issues that made it difficult for him to walk and breathe at high altitude, so they decided to stay with the pickups and take any deer that might escape down the canyon.

The rest of us fanned out and started walking up the canyon toward the deer.

I noticed that we were attacking the hill in a classic dismounted infantry maneuver. We disembarked at the line of departure and assaulted the hill by advancing in line. My position was to follow a finger ridge on the east side of the objective. I knew I had longer to travel, so I had to hurry as I moved up the ridge. As I came to a clearing, I spotted movement in the brush ahead of me.

I dropped to my knees and took cover. The targets broke from the brush and started to cross the ridge single file. There were two males and three females. I could tell they hadn't seen me. They were moving easily but purposefully away from the slope where the rest of our unit was approaching.

I pulled down on a male that was standing broadside to me. I placed the crosshairs of my scope just below his eyes. He had turned his head to look back down the trail they had taken, alert to the danger we presented. But so far, he was not aware of my presence.

I decided I didn't have a clear enough shot at his head. There were branches of brush in the way, so I moved the crosshairs back along his neck to his shoulder. Then I moved the crosshairs down behind his shoulder to his rib cage. A heart and lungs shot would be clean and neat.

Fuck you, Tom Hayden, I thought.

I moved the crosshairs of the scope to the deer's stomach and pulled the trigger. I didn't hear or feel the gun go off, but I saw his legs buckle as he stumbled and staggered back down the path from where he'd just come.

For an instant, I swear I saw that dead NVA soldier from Cu Chi standing right there, gut shot, eyes blazing at me, screaming *Sin loi, motherfucker!*

I squeezed my eyes shut and lowered my rifle. When I opened my eyes again, I saw the other male and three females standing there in shock. I immediately pulled down on the second male, before he could do something.

I was just in time to see the brim of his hat come up and expose the surprised face and eyes as they looked at me. I pulled the trigger again and he went down, he just kind of sat back and rolled over.

In one quick, synchronized leap, the three females disappeared into the brush.

Goddamn, I thought. *What a way to fight a war.*

Everything went white. Slowly, I became aware of pain in my hands. I looked down and saw both hands planted in the white snow. I was on my hands and knees, crying. My rifle lay several feet away in some buck brush. My hands were so cold I could not move them. I wanted to cut them off.

From somewhere off to my right and below me, I heard Rob yell, "Goddamn John, you gut shot the son-of-a-bitch!"

The deer hunt wasn't over for me. Later in the day I would take other shots and I would kill more deer. When the day was over, I had four deer carcasses in the back of the truck.

But after that day I would never shoot to kill again.

The excitement of the day wore off as we settled into camp that night, and after we cooked supper, I relaxed. Dad, Uncle Darold and I started telling old family stories as we drank a couple of beers.

I realized I had to take a crap. Uncle Darold had brought along a brand new portable fold-up potty. It was still in the box. I decided it was time to try the new state-of-the-art human waste disposal system. There was still snow on the ground, but the ground under the snow had not frozen yet. I took several minutes to get the potty set up, then I stripped off my pants and long johns, exposed my bare butt to the elements, and sat down to do my duty. The sensation of touching the cold potty with my bare skin caused me to pucker up just a little. But I relaxed as the cold potty warmed up.

The moon was breaking over the ridge to the east and the stars filled the sky. It was beautiful and peaceful. Off in the distance I heard a coyote call and another one answer. I was content, so much so that I could not feel the cold. My body remained still, but the rest of me floated off to meet the wonder of it all.

Then one leg of the potty sank into the soft, wet ground, and the potty collapsed. I found myself sitting on the cold, cold snow with my bare, bare butt.

"Goddamn, son of a bitch, Jesus Christ, *shit!*"

As I struggled to get back up without getting anything unpleasant on me, I heard dad call out from the camper.

"You all right out there?"

"Uncle Darold's goddamn fancy potty collapsed on me!" I said.

He and Uncle Darold burst out laughing. After I thought about it a second or so, I started laughing, too.

Dad said, "Don't you bring any of that stuff in here."

"Why the hell not?" I said. "I need to do something special for Uncle Darold."

All in all, it had been a good day.

I was lucky.

Home
Late fall, early winter 1970

By November, I was working as a laborer at a construction company. I'd been an officer and a gentleman just three months before, but now I was scum. Before, I'd made enough money to be comfortable. Now I was making minimum wage. Before, I'd been in charge. Now I was a flunky. It hurt, but I had a family to support and I'd do anything to fulfill my obligations.

We spent the winter building fences around a refinery and building a rest area on the new interstate highway. That winter was exceptionally cold, even for Wyoming. We spent the worst part of the year outside. The cold, wind-blown snow dug into our faces as we tried to complete the assigned task. We spent hours on the end of jackhammers or picks and shovels. We set fence posts with hand drivers until our arms felt like they would fall off. I came home exhausted each night, but I felt good. The hard work turned out to be a kind of therapy. Day after day I could vent my frustrations on things that were not alive. My life wasn't what I wanted, but it was acceptable; for that time and that place, the work was what I needed.

After the day was over, after I had shed my work clothes and cleaned up, my beautiful little daughter would come and sit on my lap. She would hold my face in her tiny, delicate hands, look into my eyes, and say, "Daddy, I love you."

She did not know me as I'd been before. She loved me for what I was right then. There was no confusion in our relationship. Her reality was what she held in her tiny little hands and in the eyes that looked back at her and in those big arms that would protect her and comfort her. She had no delusions about her daddy. I was what she wanted me to be, so she did nothing to change me. She was as pure as that little girl with no legs who had saved my soul in another time, another place.

I was lucky. For those who really loved me in those first critical months, an aura of immortality in my heart was their reward. They opened their arms and their hearts and extended their protection as if to wrap me in a cocoon. Patiently, they waited for my evolution and rebirth. In standing by me, they performed a sacred act. It was a ritual that had been performed over and over, generation after generation, and they did it as well as it could have been done.

They welcomed me home.

Home
Spring 1971

There were several months when survival was the objective, to meet each day and take what came my way. Slowly, concerns common to most Americans began to manifest as my own. Being a husband and father required me to become aware of our economic reality. My little family was surviving on two minimum wage incomes. In time, I realized it was not enough; I wanted more. A spark of life was rekindled.

One day, I got a package in the mail from the Army. It was my second Bronze Star. I had no idea why I got it, but it stirred something inside. The call of the bugle remained strong in my heart. There was a National Guard artillery unit in town, so I inquired if there were any vacancies. There were. They needed a battery commander, an artillery-qualified captain. It helped that I was combat qualified. So, I joined the Wyoming National Guard for two reasons. First because I missed the uniform, and second because we needed the money.

In 1971, the National Guard was a world unto itself, an interesting association of persons. There were young guys still enamored by all things military, bureaucrats protecting their jobs, older ones who were protecting their retirement, draft dodgers who were protecting their asses, and folks like me who were doing it for the money.

I expected that I would spend my once-a-month weekend drills training to be an artillery officer, a combat-qualified soldier in service of his country. I expected that once a year, I would spend two weeks testing and relearning critical combat skills.

What I found was completely unexpected. The National Guard was like nothing I had ever experienced. My weekend drills were not designed to train the officer corps in combat skills. They were designed to train us in bureaucratic propaganda. I had to learn how to generate the right statistics to justify my present budget and to justify a larger budget next year. My goal was not to prepare my battery for combat, my goal was to justify money.

I had been down this road before. I remembered my days as motor officer in Vietnam. I had failed at that job. Question was, had I learned my lesson? Could I learn to lie with statistics? Could I figure out how to justify the budget?

It fell to the officer corps of the National Guard to gather statistics. As a member of that corps, I soon learned that the real mission of the National Guard was its own survival. As battery commander, if I could justify a budget at least at the present level, I could justify the battery. The battery would justify the battalion located in Caspar, and the battalion would justify the state headquarters in Cheyenne.

The mayor and the city council were happy I maintained and spent a budget in Rawlins. My wife was happy I maintained a budget when she went to the store. Hell, it was all good. So I became a weekend bureaucrat.

The training during these weekend drills generally consisted of non-functional activities. Officers did no training themselves; they supervised training for the enlisted ranks. The enlisted ranks worked at basic training, such things as using and maintaining their assigned equipment without breaking it. But they did not train on those skills that would make them effective in combat. The units reached and maintained a combat skill level just slightly above that of the average Chamber of Commerce.

There was an unspoken understanding that the National Guard would not be called on for combat in the foreseeable future. Most if not all of the members of the unit knew their chances to see combat were nonexistent. Under the circumstances, it was no wonder that their motivation was not those of a unit of warriors.

An officer's career advancement in the National Guard was determined by his ability to recruit and retain members. Efficiency reports did not reflect technical competence in combat skills as much as technical ability to support the right statistics. The right statistics meant the right number of warm bodies, the right mix of ethnic representation, and enough of the right kind of written reports to satisfy the Inspector General once a year.

I accepted the role because it had some benefits. The biggest benefit was that for a day or so each month, it got me away from my minimum-wage job and my minimally effective marriage. Since there was really nothing to train for, we did a lot of playing.

One weekend a year we "qualified with our assigned weapons." That meant we would take our M-16s or our .45s out to the target range and shoot at paper targets. It helped us sharpen up for hunting season, and that was a good thing. I laughed a little as I watched them shoot at paper targets and thanked God that I would not have to go to combat with that bunch.

Once we demonstrated we could shoot paper targets, it was time to demonstrate we could shoot artillery. The artillery test used old car bodies as targets that were also not allowed to shoot back. It was supposed to statistically certify that we could survive in combat, and that's exactly what it did. Statistically. But it did not certify that we could survive in combat.

We could, with a little practice, do the moving and the shooting and the communication required to pass the artillery test.

We passed the tests. Flunk the tests, the budget gets cut. Pass the tests, the budget survives. The bureaucrats could justify their jobs. We could justify partying on the weekend. Failure would not do. All it took was a check mark in the yes box and everyone was happy.

It was a little demoralizing, but it was okay. If I wanted to learn to be a combat officer, I could learn on my own. And the paycheck was good, even though I was never sure when it would show up.

There were other benefits. Our summer camps were close to Fort Laramie. Names like Crook, Sheridan, Sioux, Cheyenne, and the Black Hills were at Fort Laramie. Sitting Bull, Crazy Horse, Red Cloud and Black Elk spent time there. They are remembered for Custer's Last Stand and Fetterman's Massacre. The Indians fought over the same land we played guns in, and I could see why. It was beautiful land. I'd have fought for this land, too. Knowing the history of where I stood made me feel like I was in good company.

Summer camp got me away from the job and the wife and let me dream a little. I liked summer camp.

In the evenings after we finished shooting artillery, we had time to ourselves. I used the time to explore the old Indian campsites and tepee rings and to look for arrowheads and other relics of civilizations past. It was a peaceful time for me, a time to rejoin my spiritual world.

One evening I was standing on a bluff that opened to the north and east. I could see Laramie Peak to the west, and to the east, I could see a great distance. It was springtime; the new grass and flowers covered the plains before me. The greens of the trees were new and vibrant. In the western sky between Laramie Peak and the bluff, I saw a gathering thunderstorm. I saw distant flashes of lightning and heard faint thunder claps. The sun was behind a cloud and above Laramie Peak, creating a halo around the thunderhead. I could see rain pouring from the cloud in large wide, gray streaks.

It was a beautiful moment. From someplace in a confused and forgotten past came the words of an old Indian prayer I'd once read in a book. In my mind those words changed and became personal for me.

I turned to the north, lifted my face and my hands to the sky and spoke my prayer.

"It is that I can see you, my brother of the North. I understand and honor that from your quadrant comes the purity of the white and killing winds. It is from your heart that comes the purification of my world. It is from your soul that comes the winter's cold in my preparations of my new beginning."

Then I turned to the east and continued the prayer.

"It is that I can see you, my father of the East. I understand and honor that from your quadrant comes the light of truth. It is from your heart that you send me the sun. From its light I can see all that is around me. I can see my grandmother. It is from your heart that you send me the moon, from its light I can see all that is over my head. I can see my grandfather. It is from your soul that comes the knowledge of light in my preparation for my new understanding."

And then, looking south, I prayed, "It is that I can see you, my sister of the South. I understand and honor that from your quadrant comes the breeze of new life. It is from your heart that you send me the warmth that brings the eagle, the two-leggeds and the four-leggeds. It is from your soul that comes the promise of my future."

Finally, I turned to the west. "It is that I can see you, my mother of the West. I understand and honor that from your quadrant comes the thunder beings, the power of life. It is from your heart that comes the lightning, the visible demonstration of the purpose of being. But it is from your soul that comes the raindrop, your promise to me that God has not forsaken me.

"It is but one, that is but three, that is but me. Thank you."

I could not help but feel the power, the presence of those who had passed over this land before me. This was sacred ground, and now I was a part of it, this former home of a civilization long since gone. They had come to this place to treat and to war. They had watched from this same bluff as promises proved worthless. They had watched as their way of life was forever altered. They had watched as a storm from the east had engulfed them, destroyed the foundation of their hopes and dreams. Yet from this bluff they had persevered even when they were treated as subhuman. They had, from this bluff, felt the feelings of extreme despair.

Could it be possible that on a spring day in 1875, a young Army officer had stood on this bluff, at this very spot, and grieved for the passing of the Sioux and Cheyenne nations? Could it be possible that in 1975, as I stood on this bluff and grieved for the passing of South Vietnam, I was completing a circle of life that bound time with time in a never ending flow?

Listen, my friend, there is a reason.

It was here that I realized the real cost of Vietnam. My marriage of convenience wasn't working. Vietnam had sent my wife down a road of rebellion against the chains of traditional relationships for husband and wife. Vietnam sent me down the road of a damaged soul and a desperate need for those traditional relationships of husband and wife. The friction caused by these opposing realities was unacceptable to me, and I began to realize that my love for Anna could possibly hurt her.

Vietnam cost me Anna. If I was to get a chance to end my little war, the price I would have to pay was the very relationship that had held me to life. My daughter had to lose her daddy. From that time forward, I would be required to love her quietly, from afar, and only in my heart.

I doubt that she will ever understand.

Once I made that decision, I was alone, free to take on the rat.

Lawton, Oklahoma
Easter 1975

Together we had watched and listened to the heartbreaking story of the evacuation of Da Nang. We had watched and listened to the story unfold as Captain Thinh and Staff Sergeant Son staged a brilliant defense of Zuan Loc. Zuan Loc was their finest hour, but it was too little, too late.

We had watched and listened to the story of the evacuation of the American Embassy in Saigon. That was the end of South Vietnam. The sound of the Huey helicopters and the pictures of them evacuating the Embassy in Saigon broke my spirit.

Oh, what a terrible waste. What a terrible lesson to learn. We were watching the results of America's attempt to play at war. The death beats of a mortally wounded democracy came from an American helicopter as it lifted off the roof of the embassy and flew away. The powerful *whop, whop, whop* of the rotor blades diminished then faded to nothing. Goodbye, South Vietnam.

What had my friends done to deserve this? Every time I think about Easter 1975, I think:

My friend, what a price we had to pay for your ATM.

How could this have happened? I was confused at how America could turn her back on South Vietnam, her friend and ally. How could America turn her back on the millions of her sons and daughters who had sacrificed beyond the call of duty? Because of the things that Eisenhower, Kennedy, Johnson, Nixon, and countless others had promised to them, they pledged their lives, their fortunes, and their sacred honor to an idea of South Vietnam.

How could America, my country, be so ungrateful?

I had believed that something would be done. I believed it to the very last. I had delegated that power to my representative government, and they had wasted it. I was crushed because I realized that there was much more at stake than the backstabbing of a friend.

It was bad enough that close, dear friends were no more because of a place called Xuan Loc. It was bad enough that their families, so full of hope and excitement such a short time before, were to be no more. It was bad enough that there were tens of thousands of Americans who were no more, and millions of Americans who were permanently scarred and still fighting their own little wars. It was bad enough that there were millions of American families who withstood the trauma of a loved one's bloody experience in combat. What was beyond the immorality of it all was the fact that my government had turned its back on all of these acts of faith. Now we were living Easter 1975.

An inexcusable abuse of power.

And then SSG Son spoke to me.

It appears, I am sorry to say, that your friend, and my father, Mr. Son and my grandfather, were right all along. It was all just American propaganda.

How can you apologize to us now? We cannot hear you now. To be honest with you, I wouldn't listen even if you could make me hear.

Your country is uncivilized.

With the waste of our blood you may grow wiser. If not, you too will perish.

Look at your hands and watch our blood.

Drip, drip, drip.

In the night of Easter, 1975, I cried. Teresa held me in her arms as I repeated, "I am sorry, I am so sorry."

In the fog and the noise of my grief, I kept remembering a time past of dreams turned into nightmares, of knights and white chargers, of demons and dragons. I saw the young soldier with burning eyes and a little girl with loving eyes; I saw rats and cockroaches. I saw mothers with hate in their eyes and mothers with love in their eyes. I saw fathers with the fire of patriotism and democracy burning in their hearts, and I saw dead fathers.

I saw the devil.

And through it all, I kept hearing, *Ask not what your country can do for you, ask what you can do for your country.*

I'd given my country my soul. But it wasn't enough to save SSG Son.

Teresa held me through those desperate days. She grieved because I grieved. I could not tell her why I hurt so, because I didn't have the words to express the pain I felt. She had not heard my stories, because I was afraid to tell her. I was afraid that if I started letting them out, they would destroy me. The grip I had on my world was not much, but it was all that I had.

She held me as I shook at the sound of a Huey flying overhead. She held me as I cussed at the stories about the good reasons for leaving Vietnam, or the wonderful news that Carter had pardoned the draft dodgers and deserters or the self-righteous, sanctimonious "I told you so" assholes on the television. She held me at night when I was so tired that all I could do was cry. In doing so, she created a bond that went beyond my limits of understanding.

With her delicate and beautiful hands, she gently raised my face until my eyes met hers. "Please, John, don't look at your hands," she said. "Look at me. Let's walk this walk together."

I was too preoccupied with my own pain to understand what this woman was doing for me as she held me. I did not know at the time that her father was a veteran of WWII and her brother had served two tours in Vietnam and he had left part of his eye sight and part of his soul there. I did not know at the time that she had spent her life living with and loving her men of war. I did not know that she was welcoming me into a family whose women were every bit the combat veterans that their men were. I did not realize that she knew how to love me.

I did know that this woman loved me and I sensed that, somehow, I was lucky. It took years for me to understand that this woman was and is my guardian angel. She came to me at the perfect time, in the perfect way and at the perfect place. It took many more years for me to realize that I came to her because she needed me as much as I needed her. Her life had been prepared for me and my life had been prepared for her.

Our life together had been arranged. If there is no God, I sure would like to know what has been so good to me.

You see, my friend, this is the reason you lived.

Postludes: Summer 2003, October 2010, December 2015

I pull these writings out of their box from time to time to read them. I have a different reaction each time. I am glad that in 1985 I decided to not use real names or unit designations in some instances. In my memory, SSG Son and CPT Thinh are all that need to be real to me.

This thing has turned out to be something of a diary for me. At the urging of my wife, Teresa, I started it in 1985 to deal with issues that began in 1970 with roots dating back to 1944. The issues are still here. Well, maybe some are gone, but many are not.

My wife and my number-three son, Charles—he's a writer—have conspired to have me read what I wrote those many years ago. 1970, 1985, 2003, 2010. By reading and reliving these issues I am beginning to realize what it means to get old. The process is good and not so good. The good is the intensity of some of the feelings has diminished. It is good that I have not forgotten my feelings for some very important people in my life.

The not-so-good is that time has a way of confusing the truth. Time has allowed me to think and rethink, and in so doing, the clarity of the truth sometimes gets hazy. Maybe the aging process was necessary so that I could mend.

—

I have witnessed my grandparents' generation grow old, reach its physical end, and perish as memories fade. My parents' generation is old and at the point of ending. I have come to believe that my role in this process is to live as long and well as I can. I am their memory, in body and soul.

What is amazing about this process is its perfect symmetry. Kids and grandkids, so obvious, so simple and so perfect.

My reactions to reading my memoirs have evolved. First I was hostile, mad and scared when I wrote that stuff back in 1985.

But since then, things have changed. Life has been good. For one, I have eight grandkids. The concept of grandkids is cool. Whoever came up with that concept should be commended.

Grandkids are the best-kept secret in the world. They're the reason for living, the reward for a good and productive life. They are gifts from one generation to another.

The joy I feel when I see, hear, touch, or think about my grandkids is beyond measure.

My second reaction is that, while I fought in Southeast Asia, the next generation is fighting in the Middle East. Funny thing is, they are fighting for basically the same reasons we did, trying to accomplish the same goals. The only real difference is the look of the devil and the feel of Hell.

It concerns me that the same monkeys, the ones at the roadside stands advertising drugs for sale, the monkeys who played with and fucked up my war are still at it today. It is exasperating that they work so hard.

I believe that debate of war is necessary and critical for our survival as a nation. However, if Vietnam had value to me, the value I would hope to be realized is that broken promises are costly. That promises, when made, must be kept. The smart thing to do is to think before you promise, because once that promise is made, then debate over whether it was the right thing to do should be delayed until we've lived up to our promise.

If we are truly world leaders, if our vision of good and evil has merit, then there should be no higher goal than to make and keep promises.

There is another reaction. Vietnam does not mean the same thing to me now. I don't think about it as much. I don't think about it the same way. Actually, when I think about me and Vietnam, it seems like a tall tale. There is a real fuzzy line between truth and fantasy. I have had too many dreams, too many nightmares, and too many daydreams to be able to sit here and say, "That was the way it was." I don't know for sure what's true and what isn't anymore. I guess, in a way, it's all true. But in another way, none of it is.

Three things are certain: in February 1970, I flew in a United Airlines Boeing 707 to Vietnam. In September 1970, I left Vietnam in a United Airlines Boeing 707. And I am sure I was a captain in the U.S. Army the whole time I was there.

—

Here I am, still at this. The process has taken thirty years to write and fifty years to process. Teresa and my son Charles are pushing me to finish this thing. They want to get it ready to publish. I have no idea why they would want me to do such a thing.

I have to confess that I really don't like reading and editing what I wrote those many years ago. I really hoped I was through with those old dreams and those old emotions. I guess I am not ready to put it all to rest—not yet.

It makes my heart ache to think about SSG Son, CPT Thinh, and my daughter Anna. I lost Son and Thinh because I failed to take up arms in their hour of need. I lost Anna because I just walked away from her.

That they are not yet forgotten is the best I can do. They gave me so much.

Maybe there is something of a memorial, at least in my mind. Maybe the reason I am still here is to remember them in my own way. Maybe this is how forgiveness happens.

There is not a lot of pride in my heart for this part of me.

But for the rest of me, there is pride in my heart. The last forty-five years have not been a failure. I am a proud grandpa, a proud father, and a proud husband.

My guardian angel, Teresa, remains the strong and intelligent manager of my soul. We have been married a long time, and she remains the beautiful, sexy lady on my arm. Something to be real proud of. I am still amazed that she does what she does for me, in spite of all my shortcomings. I hope she realizes that I appreciate the opportunity to be part of her world.

And there are my three boys. Each one is a success in his own right. They built and are still building unique worlds for themselves and theirs. That's something to be real proud of, too.

They do not shirk their duties and responsibilities as head of their households. They let me be their father and a grandfather. What a great feeling of pride for an old man.

I have eight grandkids! What a rush. Each of them brings to me a story and a life that is so unique and beautiful. To look into their eyes, hug them, and listen to their take on life, knowing that I am a small part of their world, is a privilege. A privilege that sometimes I think I don't deserve.

But then, what if . . .

SSG Son? CPT Thinh? Would it be OK with you if I offer, through myself, my family to you? I think, maybe you are in my grandchildren, the good part of my grandchildren.

—

When thinking of war, specifically my little war, I remember someone said something like "To the winner go the spoils." I also remember someone claiming that the only real spoils from winning a war are the rights to define good and evil, right and wrong, and truth and fiction.

Since I am still here, I am going to declare that I have won my little war.

Which brings up an interesting question.

What have I won? Maybe I have won the right to tell the truth as I want it to be. I do not doubt that the argument I had with Twinky on New Year's Eve did not come off exactly as I storied here. It is also true that it came off exactly as I want it to be now.

Isn't the passage of time a marvelous thing?

There are some things even a winner can't change.

"A nation that draws too broad a difference between its scholars and its warriors has its thinking done by cowards and its fighting done by fools." —Thucydides